IN THE RING

Dreamers, Book 3
J.K. Stephens

Daybreak Publications

In the Ring, Copyright© 2020 by J.K. Stephens

Published by Daybreak Publications

www.daybreakcreate.com

email: jKStephens@daybreakcreate.com

All rights reserved. No part of this book may be reproduced, stored in a retrieval system, or transmitted in any form or by any means, electronic or mechanical, including photocopying, recording, or by any information storage and retrieval system, without the written permission of the author, except where permitted by law.

This is a work of fiction. The characters, incidents and dialogue are drawn from the author's imagination and are not to be construed as factual. Any resemblance to actual events or to persons living or dead is entirely coincidental.

ISBN 978-1-7328660-6-5

E-reader ISBN 978-1-7328660-7-2

First Edition. Cover design by JD Smith Design

To Angel, who taught me that the song is everything.

IN THE RING

Dreamers, Book 3
J.K. Stephens

CHAPTER ONE

FREYA TURNED IN HER SLEEP and pulled a handful of curls out of her face, pushing it back under her head against the pillow. She dozed again on the edge of waking, and a dream came to her:

The cats arrived in twos and threes, tired but eager, moving out of woodland shadows into the daylight.

News that the legendary Singer had come drew mountain and desert cats from every direction to the gypsy camp to observe and report.

"She is weak," a cat reported to its Gathering, and Gatherings reported to other Gatherings, and the next Gatherings reported the news on and on, in chains of news that reached places far from New Mexico.

The dozens of visiting cats caught fish for the Singer, brought her herbs and fresh wild rabbit.

"She can't eat. She has no body," the very old human reminded them. But the gypsies cooked the

offered food for her to breathe for sustenance, then shared some raw and some cooked with the visitors, to show them a return of courtesy from the Singer.

The Singer said the vapor of the cooked food reminded her of home; how long had it been since she had enjoyed such a meal?

"Get a body!" the cats urged, full of practical ideas.

"No time for that," she answered them, just as practically.

**

They slowed the horses at the top of the rise and Freya mimicked Ozzie, turning Cheetah's head as Ozzie reined Rory around to face the valley. Cool air from over the rise ruffled Freya's hair, pulling wisps of gingery stuff out of the bun she had made to tame it. Cool wind made her freckled cheekbones pink. He saw her again in Egypt: stunning, she was so beautiful. And saw her as she was on Mars, staring stubbornly at him: fierce and mean and awful. Right now she was neither, he noticed.

The horse sweat and leather smells were familiar and comforting to Ozzie. He wondered if she minded them, but she said nothing.

She didn't ride like it was only her second time. The first had just been a few months ago, in June, on his father's horseback tour of the campground. Just three months ago. But it seemed like years between then and now, with Mars and their impossible, miserable, triumphant search for the Singer sandwiched into the middle.

He looked out over the valley, listening to the cicadas rasping and chattering. How sleepy this place had seemed to him just a year or so ago! Then, it was the Nothing-Happening Department of the whole universe.

Freya sat with her hands clasping the reins and resting on the pommel of her saddle. She looked out toward Space City—Las Cruces. He watched her watch the clouds go pink, then golden and white as the sun cleared the tops of the Jarilla Mountains and became too bright to look at directly.

She breathed deeply. "So quiet up here," she said.

"Uhuh. If you listen hard you can hear the noise and traffic from Las Cruces here sometimes, but not this early in the day."

"And that's the spaceport?" She pointed at a glob of still-bright lights in the distant haze north of the city.

Hard to believe, in this calm, that it was just a week since they had exited from the Mars Science ship and run into the desert darkness from that spaceport, barely managing to remain undiscovered as stowaways. "Yeah. Way out there, just east of the big highway," he said. "See the runway lights, like a squashed star?"

She nodded, and then turned to smile at him, showering cool blue sparks right into his eyes.

He would never get over that. He said quickly, "I wanted to show you how pretty it is at dawn."

"It *is* pretty," she agreed. "I love it here."

He could tell that she was restless, though. The truth was that he was too, and just when they all probably needed the opposite—a little pressure-off time after the exhaustion of Mars.

Norm and Alexis were probably still sleeping right now, way below in the old house next to the campground. Where Dad and his new wife Ilse (Surprise, Ozzie! Freya's mother is now your stepmother) were probably asleep too after a late night at a singing engagement.

The morning wind through the pinyon pines behind them carried a roaring sound, like flame. His breath caught and he turned his head quickly at the sound of thunder. Or the earth shaking. Coming from these squat, placid hills? Impossible.

Still, he felt doomed.

He'd felt this way before. *I guess this time I don't have to wonder why.* Between the volcanoes and an enemy or two, there was enough reason to be uneasy.

In spite of opposition, he was eligible again for the Captain's Apprentice spot on the Grand Galactic ship *Liberty*, departing for its first-ever Earth-Moon-Mars trading flight in January. That was all possible again. How likely, though, that Raker would forget about them after they had made an ass of him and exposed Seth? The

Representative from Pasadena and his son had not forgotten them, Ozzie was sure.

What's Raker's problem, anyway, that he couldn't leave Freya and me alone on Mars? This can't be all about a space-trader apprenticeship. Or about a clever little robotic greenhouse on Mars, with chickens and tomatoes. The guy seems to be a little frantic about stopping us— keeping us from doing something. Why?

He nodded at Freya then, and turned Rory's head toward the trail again. He could hear Cheetah's breathing and feel the big guy's warm breaths coming from behind him on the narrow track. As Rory picked her way downhill, around yucca and cactus and rock, he thought out loud: "We'll see my mother again today. She wants to talk with us about the Singer…" But Freya already knew that, of course.

Almost daily for the last week they had been going to the campground, where Ahanith, the Singer, was with Malo for safety. There she practice-sang with the gypsies and warmed up to give it her best when they took her to record and test at the studio in Las Cruces.

And they checked the newsfeeds each day. While they had been on Mars, no new volcanoes had started and no old ones had come alive. Nothing new since their return a week ago, either. That was good news.

But now that they had brought the Singer to Earth, there should be big changes for the better, shouldn't there? The past week had shown only a slight reduction in volcanic activity everywhere. And no big improvement that they could find, anywhere. She'd been singing for hours, daily, for a week.

What if she didn't have enough horsepower for Earth, a bigger planet? Maybe Earth was more messed-up than Mars?

"Maybe your mother—Diana—will have an idea about what the problem is," Freya said. His mother's environmental investigation company, Premier Independent Investigators, had a record of unusual research that contradicted its ordinary-sounding name. PII had taken this on as a project. And Freya seemed totally won over by his mother, who treated Ozzie and his friends as research associates and insisted that they call her by her first name.

Rory balked for a moment at a cat that leaped across the path and disappeared among the yucca on the downhill side. Behind him, Cheetah blew out a breath.

Freya said, "Here's some news I didn't tell you yet. Not a surprise, though. The 14th Reykjavik Fire Company is still mad that I left for Mars, of course. When the Captain asked what our work on Mars had to do with the volcanoes, Diana said it might be best not to make it all public at this time—and I believe her—so I told him that we went there to do research into vibrations. Truth. But he said it was hard to believe as a reason. He wanted more of an explanation.

"What could I say? I just told him, 'I'm fighting the

fires in the best way I know.' Silence after that. So. There goes the job I think."

Well, she *was* still fighting the fires. It just wasn't obvious yet. It would be, when the Singer's amazing voice-power started to make a difference. But he knew how she felt: hard on your pride to be kicked off a job.

It was probably good that Ilse and Dad would be away for part of today so Ozzie and Freya could talk with Diana freely. There was too much that Dad and Ilse didn't know, and maybe didn't need to...

Partly just to see what Freya would say, he asked, "Did your mother tell you? About my dad going with you two to Iceland?"

She was silent a moment behind him. "Yes."

The horses took Rory's shortcut and slid gently on their haunches, like skiers, downhill on the loose sand and rock between two loops of the trail. He had forgotten to warn Freya, but she pursed her lips, held on, and slid beside him.

Yesterday Ozzie's father had announced that he and Ilse had new plans. They wanted to do more playing locally, and in fact they were being called about bookings in Phoenix and Flagstaff next, and other cities "all along the SpaceRail to L.A." Dad was proud of that. But months ago when she needed the work, Ilse had committed to teaching art in Iceland for the coming year, and she would keep her promise. It was time for her to return to Reykjavik.

So the latest was that their newlywed parents would go together to Iceland. Volcanoes and all. They had decided to get themselves booked to sing together there. Today they would talk with the band members to plan a return tour in New Mexico next summer...

Dad, leaving for Iceland. Freya, too. But what would she do there, with no job?

Maybe he would miss her. He guessed he would.

OK, time to get real: I will miss her, a lot.

But he had to stay here to nail down that apprenticeship... He had done what he promised her and more. *The Singer is here. I managed to help Freya get her here. Don't know what Freya wants to do next, but I know what I need to do.*

Hard to see what was beyond that.

A cicada hummed loudly behind him. It sliced past Ozzie's ear, an insect bigger than his thumb, with a blur of whirring wings. "Ozzie, look out!" Freya's voice hissed behind him.

The thing had turned and was arcing back toward his face. It was *not* a cicada.

"Get down, Freya!" He pulled his old pistol from inside his jacket and fired without thinking as he ducked, up into the middle of the nasty-looking device. Whirling blades and sparks and ozone flew in all directions: high into the air, into his hair and face, and behind him at Freya. The horses shied, screaming with fear. Landing bits lit the sagebrush.

IN THE RING

Ozzie grabbed at Freya's reins, which were whipping from side to side as Cheetah's head jerked back and forth. Ozzie's hand missed them but Freya lay forward onto Cheetah's neck and snatched at them herself, snagging one in each hand, and pulled them up hard to tell the horse to stop.

Ozzie couldn't help a grim smile at Freya as he reined in his own bucking and skittering horse. Rory tossed her head too, but her snorting and shrieking calmed as he talked to her. It didn't help that a few cats leaped past Rory's hooves just then, yowling with fright, to disappear downhill into brush and pinyon.

What are cats doing up here?

The flames in the sagebrush crackled out one by one, leaving scorched branch-tips smoking to mingle with the smell of his own burnt hair.

When Rory and Cheetah were still again, he looked around. He and Freya regarded each other, breathing. He raked back his hair with one hand. His heart thumped less loudly in his ears. The quiet was punctuated only by insect noises again—real insects, he hoped.

He reached out to hand the gun to Freya: "Cover us; shoot if there's another one." Wondering if she would hit him by mistake, he reminded himself to do some target-practice with her soon.

He dismounted and quickly picked up all the wreckage that was left of the robot-cicada thing.

"Let's get downhill into cover," he said.

CHAPTER TWO

"WAIT, YOU'RE BLEEDING," she said before they left the barn.

They had just put Rory and Cheetah in their stalls to eat breakfast. Freya pulled a tissue from her pocket and dabbed at his forehead. He squinted and flinched a little.

"Haven't you ever had someone wipe your face?" she asked. There were a dozen little slashes, oozing, but nothing serious-looking. One of them bled into his left eyebrow and dripped to stain the thick lashes on that side. He eyed her as she wiped off the last of it.

"Guess not," he said.

She looked him over. At his wheat-colored hair, brushed back "Las Cruces style" but less sun-scorched after months in the artificial Martian greenhouse light. It had grown so long it hung below his shoulders now. Maybe later she would offer to cut it, if she had the chance.

IN THE RING

**

Norm and Alexis were on Ozzie's porch steps when he walked up from the barn with Freya. Ozzie held two handfuls of razor-sharp junk delicately. They had emptied from his saddlebags all the remains of the exploded thing, to show the others.

"Well, you two are up early," Alexis said. Despite the hour, she had her dark hair brushed neatly and woven into the usual long fishtail braid. She sat beneath a flowered green canvas sun hat that looked too cheerful for her right now, and said, "Newsfeed, Freya. Not good: two old volcanoes have gone active."

"Where?" Freya hastily took the step beside Alexis. *This is how we know if we're winning or losing.*

"Well, the volcano scene has been improving all week. Cuz the Singer is here now," Norm said. "She just needs to get her power up a little—right, folks?" A wise-guy grin lit his bronze face. He lay on the porch floorboards poking at his phone, apparently adjusting gradually to daytime by starting the day in a prone position.

It took a lot to worry Norm.

One by one Ozzie placed the pieces next to Norm on the smooth gray boards.

Norm became fully awake. He rolled over and looked.

Alexis found and showed the worldweb images on her phone: Captions said the newly-wakened volcanoes were Kilimanjaro, in eastern Africa, and Vulcano, now spitting up fresh lava near the toe of Italy.

Freya's head ached. "Look, Ozzie. Nothing in recent history in either place, and now this."

Alexis said, "How can this be happening if the Singer is here?"

Freya shook her head. *Don't know.*

"Listen," Ozzie said. "Norm says it's a surveillance device."

"Yeah," Norm said. "I've seen the designs for these on the worldweb. Old tech that seems to be making a comeback: a big-bug robot that records and transmits voice, location—" He made a sour face that distorted his perfectly-dotted freckles.

"And also, Frey, it's—"

"—fully equipped to kill. Or maim, or what have you." Norm tried his best to be funny about it, and failed.

Alexis' mouth became a startled O. Freya felt a little sick.

Voices swelled in the kitchen, and the hinge on the screen door squealed.

As if they were instantly moved by the same thought, Norm and Ozzie plucked up the singed and twisted bits of metal and pitched them into the darkness under the porch just before Ozzie's father stepped out, bearing a coffee mug.

"Left Rory saddled for you," Ozzie said.

His father waved his way past them, putting his hat

on as he went. In work jeans and old calf-high boots, he seemed to be heading for the campground to do his daily ride-through inspection early. He turned at the gate. "Today is packing day," he called to Ozzie. "I need to talk with you after the band meeting..." He waved again.

If the snooping weapon had located Ozzie, could his father also be located by surprise and injured?

Alexis held up one finger, warningly, till Doug entered the barn. Then: "Listen to this, guys, another newsfeed—"

"—Aw, I don't pay much attention to the NOOOZ," Norm yawned, pronouncing it like the HoloTube comics did.

"Me neither, but I'm still running news checks on Raker and Seth"—they all looked at her—"you know, just to see what comes up. Listen to Raker:"

> **...our proposal to the World Council is that Iceland should be taken over and managed by an international panel of experts because the Icelandic government clearly is unable to protect its citizens from a natural threat—**

"Liar!" Freya's cheekbones went hot at the sneaking injustice. This wasn't just about her, or Iceland. Someone was shooting at Ozzie now.

And after all the risks they had taken to get the Singer here, Ahanith wasn't making any difference yet?

She knew Ozzie wanted to stay in New Mexico to get his apprenticeship. Was he even safe being around here? Her eyes scoured the sky, looking for anything odd.

She had a bad feeling about Grand Galactic, "the world's greatest space trading company," because of the way they had changed their minds, back and forth, about Ozzie's apprenticeship.

She wanted to stay with Ozzie and work on the volcanoes problem from here with the Singer. But he had not invited her yet. She had invited herself here the first time, in June. Now she hoped to hear that he wanted her to stay.

After all that friction on Mars, maybe he wanted her to leave.

**

"That's right, Mom," Norm said. "It's just what I told you last week—it's all covered and I'm working on it.

"Nah, of course I won't lose my scholarship... Yeah, give Belinda and the little guys a hug for me, willya? And say hi to Pop for me... Miss your cooking! Nothing like it on Mars... OK, talk to you more soon..."

Nothing like her cooking here, either. Ozzie sat at the kitchen table, where Norm and Alexis were writing up their reports on the Mars project—for school credit, in Norm's case. He tossed more icons into the air and started another short holovid on robotic surveillance weapons.

Insane. A murderous robotic insect sent after me and Freya?

The holo image he had pulled up was shaped like a gliding bird. It hung quietly in the air, turning as indicator captions blinked on and off beside it, silently displaying its features. Even with the sound off so it wouldn't distract Norm and Alexis, it was a relief to have this worldweb connection again, after Mars. The image was fascinating. *But crazy.* It made his skin crawl. A week ago, they had almost been kidnapped. Now this thing.

His hand sought the heavy gold ring that hung at his collarbone from a chain. The metal felt hot again. It seemed to be happening more often these days.

Robot-insects could be aimed at them all—Dad and Ilse too. But if they told their parents about the danger, they'd have to answer a lot of questions.

"Ozzie, tell her that she could get into Berkford in a heartbeat."

"Huh? Of course you could, Alexis. You kidding?" Then he thought to ask: "Where exactly *is* Berkford, anyway?"

Norm explained: Berkford, the joining of two big former universities, had multiple campuses sprawled around the San Francisco Bay area. Ozzie had never been there. He'd go visit Norm sometime, after his apprenticeship. To celebrate.

Ozzie looked, and located Freya lying on the couch in

the living room under an open window, reading an old pulp-book she had probably pulled from the nearby shelf. The late August air was hot and dry, and somehow that made the sound of the magpies and jays squawking outside the window seem to be coming from all around her. He liked seeing her there, looking like she was at home and at rest, even though it wouldn't be for long.

Alexis and Norm went on arguing cheerfully about their educations while they punched in their reports and attached documents to them. They seemed to have forgotten the killer-cicada and Raker already. School existed in a different world for them, he guessed. Probably a safer one.

Ozzie turned again to the holo, but he said, "So you're late for school, Norm?"

"A little. They were expecting me last week, according to the notice…" He shrugged, grinned at Ozzie, and kept on punching at his phone.

Alexis said, "He's already got approval on his proposal from way back in June. To be forgiven for being late because he was on a Mars robotic project that could count for credits… And now he's attaching the commendation from His Majesty's government, right, Norm? How can they refuse, Ozzie?"

Ozzie nodded. "Your butt is covered again, Norm."

"For first semester." Norm shrugged. "At least I won't get it scorched for being late. I want to stay here and look into these surveillance critters with you. Just a little while. More interesting than Freshman English..." He punched one more time, with finesse. "That's it. Sent." He gave Alexis a rebellious look.

She ignored that. "Mine's almost all written up, too," she said.

"You trying to get out of school too, Alexis? To stay here a while?" Ozzie wondered what they would all eat, on no money after Dad and Ilse (and Freya, too) were gone to Iceland. Maybe Norm and Alexis weren't thinking of that yet. Well, there was the tail end of the garden vegetables, and eggs and milk, and he could hunt some small game...

"No, I'm just writing up our greenhouse study, so I can send it with the robot system plans and all the other electronic documents from the greenhouse. It's what they paid for when they bought the study, you know? I owe them that.

"But I *will* need to do something to get a scholarship. Got to have a scholarship to go to university. It's so late; maybe I'll start in January..." She looked at Norm, who was already submerging in a holo game.

Of course: she wants to be where Norm is. That's why she's stalling.

Over there on the couch Freya turned a page, then turned her head and raised her eyebrows at him. He got her message, loud and clear: Anyone here have time for the really pressing problem?

Alexis' phone sounded. She bit her lip and took the call, raising one small finger at Norm to shush him. "Hello, Father. Yes, I'm just finishing my report to send off to—

"—They will probably answer my application soon—

"—I know, I'm just looking into that now; it's been such a fast-paced summer—

"—Well, I'll bet I can get scholarship money so you don't *have* to pay for it—

"—Not much longer, I think. We're... wrapping up here. Don't want to be rude to my host, of course—

"—I will, and how are the—

"—That's good. Let me find out more and call you again."

She sighed and set her phone down quietly, with a hunted look. "They treat me like I'm a little kid," she said. "Worse, sometimes. Like I'm a puppet, or some valuable property that they own. Wish my parents were more relaxed, like my Uncle Yong. Or like your family, Norm."

Freya had arrived at the table, her finger closed in her book. She gave Alexis a sympathetic pat on the shoulder. "I'll make some food, OK? Ozzie, what do we have that I can put into the pot?"

"I'll come look and help dig things out."

Alexis and Norm looked at them, then at each other. Just looking at each other was all it took to start the two going on their debate again.

**

He had helped Freya find eggs and cheese and a little beef in the refrigerator. And in the garden, some carrots, beans, and an early squash. They put their heads together on a menu.

While he split the squash and buttered it, Ozzie thought again about the first Grand Galactic trading flight to Mars in January. He sighed with happiness, just envisioning it again. Now he could be there. That apprenticeship was *his* higher education. And his future, too, all in one package. All that he wanted.

Or a lot of it. Too bad Freya couldn't be with him during the apprenticeship. But no one could; he would have to succeed or fail alone.

Freya was always so sure of what she was going to do. Always had a plan. She would... do whatever was next for her, and now that he had a chance at a space-trading career again, maybe she really would fly with him someday, on his own trading ship.

Then the doomed feeling came back. *What about this killer-robot? What's going to happen, really? The first Mars trading flight is supposed to leave in four months, but right now, the earth is still lit up with volcanoes. And two more of them, now. Will anything **at all** be taking off in four months?*

Freya's vision of the danger was true enough. A year ago he thought her vision was vivid and scary, like a great story you could read, the kind that makes you feel bigger than you thought you were. Hard to put a book like that down, and hard when it ends.

The trouble is, her vision seems to be absolutely real.

And something else: Her vision has become my life. Actually, it's not something I can put down at all. Right now I wish it was.

CHAPTER THREE

DAD TEXTED: NOT COMING HOME till later, still talking with the band members in Las Cruces. No need to wait, so the four of them landed on the porch for lunch. *Not bad,* Ozzie thought, tasting the steaming-hot stuff. He and Freya had cooked up something that was actually better than average.

As they sat on the steps with their plates on their laps, he saw a cat ease through a gap in the gate and limp shadowless into the front yard. It was spotted black and white. Skinny and wounded too, with a torn ear and patches of dried blood matting the fur on one shoulder. It looked familiar.

It came closer, warily eyeing the water in the pump tub by the lower step. Ozzie leaned a little and inspected without getting too close. "Thought so. An old Tom," he said.

Then he realized that the habit of treating cats like the rest of the barnyard animals had made him forget, again. It just wasn't worth it, to offend cats. He inquired, politely enough: [Something to say?]

The cat stared at him. Only then did the dream from last night return to him:

A sparkling planet turned in space, moving like a stately dancer on a stage lit by the vast spotlight of the distant sun.

The graceful sphere was Earth. Or was it Mars? The sparkling came from a lighted mesh that held the planet like a fishnet. The fibers of the net were lit a little, but where they intersected they didn't just send sparks; they leaked globs of light as if the light was liquid energy, released by the knots, that couldn't be repressed.

Each knot in the net pulsed with light, while the planet turned.

He was drawn closer and closer to it.

The pulse seemed to come from some music, almost loud enough to hear...

Last night he had wakened suddenly on the lumpy fold-out couch in the living room to find this same skinny spotted cat sitting beside him, gazing at him. Norm, on the other side, had his sleeping bag pulled tightly over his head.

"How'd you get in here?" Ozzie had demanded. The thing looked like it had just been in a fight.

IN THE RING

The cat narrowed its eyes almost to slits and continued to gaze unblinking at Ozzie until Ozzie remembered: [Hello. Something to say?] He hoped the sleepy question sounded polite.

[No.] the wounded warrior leaped to the windowsill, then out through the open window again.

Now noon sun bathed the Tom, who stared at him appraisingly. The cat ended the appraisal with a loud hiss. He leaped up onto the far edge of the tub and balanced on the rim, lapping eagerly.

Ozzie's phone sounded: NO ID. That was impossible, with today's phones, but somehow her company technicians had arranged it. "It's my—it's Diana," he said to the others, setting his half-empty plate down beside him.

"When can we get the Singer out here today for recording?" she wanted to know. They set a time.

Then he asked, and she said the results from the last few days were in: a small change in the world seismic readings that her company was graphing.

"A miniscule improvement," she said.

"But since then we've got some more volcanoes leaking lava..."

"Right. Doesn't look like the Singer is an instant fix for the volcanoes, Ozzie. But that doesn't mean it's not possible. Let's talk. Hopefully soon we'll have a breakthrough."

"This is pretty good, guys," Norm rose for a refill.

"Bring the skillet out. Don't hog it all," Ozzie demanded. He walked away from the steps, down the cracked concrete walk toward the gate for quiet.

Ozzie had begun to understand his mother, and why she had chosen what she did. It took plenty of imagination not to resent her for leaving him without a word for years. You had to really think big and imagine amazing reasons for it to be OK.

*The thing is, all her reasons **are** amazing. She's magnetic, brilliant. And she keeps dishing up real-life amazing stuff.* Maybe it was strange, but he didn't hate her or anything. Most of all he was curious, now: he wondered what he would find out from her next.

She was saying, "…You know my company—PII—has found that Freya's theory about vibration and the volcanoes was mostly correct. Two days ago we released the resulting research discoveries, along all the news channels and scientific channels we could get, to encourage more and faster research. And did you see Raker's news release this morning in reply?"

He hadn't. *This morning? While he was also maybe sending killer insects after us?*

"His news release has the title "New Mexico Company Releases False Volcano Research," so you can imagine what the rest of it says. And he's got it out on all the news services, radio, holo. He seems to have a lot of friends in the big news companies."

And he sure is trying to shut people up, Ozzie thought.

Freya, me, and now PII.

"Thanks for the lunch, cooks. I need to go finish my report," Alexis said. But she and Freya had pulled out their phones and she dawdled, thumbing hers, pulling up and tossing away icons and answering messages. "Ozzie, wanted to show you one thing: look at this stuff I found on the worldweb about Dr. Tersey, your grandfather. The guy seemed to pull future technology out of his hat."

"Yeah. My—Diana—said something like that, too: she discovered that anything he came up with, she could count on to be useful as an innovation or to solve problems caused by technology—just had to find out where to apply it."

"But the formula engraved in the ring looked different to her?"

"At first. She and PII couldn't figure it out so the ring formula seemed like an exception—maybe just a keepsake or something—until she found out the formula had to do with sound."

Ozzie scraped the last sauce off his plate and licked his fork. He rose to carry plates in. Alexis stood and dusted off the seat of her jeans.

Norm said, "Hey, buckaroos, dig this *World Geo* story on Raker: He is sure one busy guy, has something to say wherever you look..." He scratched his head all over, leaving the hair sticking up in all directions.

"He talks about the 'threat of seismic disruptions' in Vulcan Arenal, Costa Rica, and what he thinks should be done about it. He's the *expert*, right? So he says Costa Rica should be taken over by a World Sciences team that should run the country...And he names other places that should have their governments replaced, like Malta."

Ozzie considered it. *Wonder how the people there will feel about being told what to do by a committee from some other part of the world?*

"Weird how these 'danger spots' that Raker is coming up with are all little places someone could push around easily, like Iceland—"

"Maybe they *think* they can—" Freya scowled at Norm.

"—or Malta or Saint Lucia..." he hastened onward, grinning annoyingly at her.

Alexis smiled a slow, lopsided smile that traveled up one side of her face. "Freya, it's a good question: why is this Raker's passion? He's *just* a lower-house rep from Pasadena."

"What do you mean, *just* from Pasadena?—But yeah," Norm allowed. "Makes no sense that this local guy is trying to run the world...

"And besides, what I was *trying* to say, Freya, was—see how much of this volcanic stuff is happening in California now, like Mt. Shasta is smoking, and look at the rest of this list in *World GEO*: six new hot springs in northern California this year, and there are tremors constantly now along the San Andreas fault—and he's not talking about anyone taking *California* over."

"True, Norm!" Alexis said. "Guys, it's true. Raker's rather oddly picky about who he decides to take over, isn't he?"

"Maybe he already took over California, and we don't know it," Freya shrugged.

Ozzie grinned, and stifled that. "You're all right in the bullseye, for sure," he said. "Something weird going on here."

Through the gloom of her face, Freya's eyes still simmered with hot sparks. She poked crossly at her phone.

Alexis took up her plate from the steps again.

"Scumbag," Freya scowled furiously up at Alexis, who froze, looking blank.

"Huh??"

Freya looked up at them all from her perch on the steps. "Listen, my friends, a just-now Raker news release: Raker gives a speech in the U.S. House of Representatives. About 'a ring of international conspirators.' It says, 'Included in the ring are four students, two of them from the U.S., and an unknown number of others, who may be armed and dangerous. They are at large in New Mexico or nearby states. Congress has authorized an investigation into their activities and connections...'"

Ozzie froze too. Alexis looked horrified. "Hah," Norm said. "Ridiculous."

It seemed that Ozzie heard rumbling coming from

somewhere. He looked overhead for a plane that might be making the noise. But the cloudless sky was so empty it looked as pure and peaceful as blown glass.

He clasped the ring on the chain at his neck. It was hot again.

CHAPTER FOUR

NORM AND ALEXIS CLATTERED in the kitchen.
Next to Freya on the porch's sun-baked lowest step, Ozzie scowled at the holo image. He used a forefinger to make it rotate, then rock away from him so he could see the underside. Freya watched, fascinated now that he had turned on the narrative voice that described the features and mechanisms inside the thing. *This is what they think they need to do to us?*

It was another one of those flying weapon-things, but to her relief this one just looked like a fat cigar—not so disguised as the others had been. *Nasty to think of birds and insects that are assassins in disguise. Twisted.*

A breeze moved the cottonwood branches overhead and shook the sunlight onto Ozzie's face. He flipped through a new holo to get an exploded view of the inside parts.

The screen door slammed behind Norm and Alexis, back already. Alexis' dark eyes were troubled. *She doesn't seem so eager to finish her work.*

Alexis sat down on the middle step behind Ozzie and Freya, drawn by the robot holo. Now Norm stared, too, at the slow-mo action of Ozzie's robot image. He watched the razor-like parts somersaulting through space. "Ouch, wouldn't want to be followed by one of those."

He jumped when his phone gave its usual explosive ring-tone. "Holotext in from Berkford! OK, here we go!

"—What?? To be confirmed by postal mail?

"Awww, dammit. Listen. It says,

Dear Mr. Garcia,

Thank you for your submitted findings and your proposal for credits.

In order to remain eligible for the awarded scholarship, you must arrive before September 15th and enroll standardly, then personally present your written proposal to your advisor. Otherwise it will be our unwelcome duty to find that you have forfeited the scholarship.

Sincerely yours…"

Ozzie made a commiserating face at Norm. Gloom grew around Norm and Alexis like gray fog.

"Sorry, friends," Freya said. She knew that Alexis and Norm wanted more time to be together here—or anywhere, really. *For me, I want us **all** to stay together. I* *think we all want that. Anyway, most of us do.*

Norm watched without joy as the mail delivery guy pulled up to the inner gate and got out of his aircar. He said meditatively, "In California, mail is dropped by drones, like the newspapers are: into roof-tubes."

"In London, too," Alexis nodded.

"In Iceland, we have a postman, like this one." Freya rose to walk to the gate with Ozzie. She needed to move.

Ozzie told her, "Here, the 'postman' is actually a message and delivery service owned by a local company. The state of New Mexico hires them. They deliver all the mail outside the Las Cruces city limits, and also all business courier envelopes and legal documents. Even flowers—"

"—And singing telegrams!" They had reached the red-cheeked mail carrier, who grinned at Ozzie. "Want to send some flowers?" He rolled his eyes toward Freya as a hint. She rolled hers back at him.

"How's business, Marty?"

"Never better. How's school?"

"Finished!"

"Good work!" He shuffled through the mail. "What's next?"

"An apprenticeship—with Grand Galactic. I hope."

He handed Ozzie a bundle of mail. "Think I saw a letter in there from them! ...I know you: you'll get whatever it is you want." He let his eyes park on Ozzie for two seconds; then they scanned the front porch, where Alexis and Norm sat. "Your Dad home?"

"No. He and Ilse left for a meeting with their band. You know he's married now. And he and Ilse, his wife—"

"Heard about that, good news. Gave 'em my best at the time. Tell him I'll stop by at the end of my run today and say hello..." He waved, squeezed his well-padded frame back into the little postal car, and was gone.

"When does Marty ever 'stop by,' except to deliver mail?" Ozzie mused, as they turned back toward the others on the porch.

Ozzie carefully opened the letter from Grand Galactic and let her read it with him. Then he crushed it slowly and angrily tossed it aside, into a breeze that bent the tops of the drying wildflowers and field-grass in the yard.

She retrieved the letter before the wind could take it away entirely.

His anger was only growing hotter. But he let her read the main part aloud so Norm and Alexis could hear it:

Because you were unwilling to show yourself on Mars, Grand Galactic believes that you were in fact a stowaway, and that you were not there as a legitimate research assistant with British approval, despite the assertions of a few British personnel.

Grand Galactic is unable to bring court action against the two reported stowaways who are claimed as assistants on the British Greenhouse project, because the only Grand Galactic property that was used by these stowaways was berthing that was already purchased by British Science.

But Grand Galactic can refuse to entertain the idea of a stowaway receiving the annual Captain's Apprentice award. Therefore your name, Oliver Reed, has been removed from the eligibility list for the 2066 award.

She knew: Grand Galactic would send copies to the Academy, probably to everyone who knew Ozzie there.

"Raker." Ozzie said. He muttered his whole string of juicy swearwords. Alexis nodded sadly, but none of them could claim to be surprised.

"Damn," Norm said. "And you need some new foul language, Ozzie... Slimy bastard probably sent that killer-cicada after you, too." Norm hawked and spat off the porch into the air, a skill he had learned from Ozzie. It landed about thirty feet away in the weeds.

"Hey, listen, buddies," he added, and grinned an evil grin. "After all his skullduggery, shouldn't we be wondering whether Raker is the one who got that trucker to try to kidnap us last week? Maybe he was trying to catch us to get proof that Alexis and I were hiding stowaways."

"Or for worse reasons." Freya shrugged, her attention riveted on Ozzie. She pushed away the picture of a prison cell that kept returning to her mind.

"What do you think, guys?" Norm persisted. "Maybe we should investigate Raker."

Ozzie shrugged at Norm in answer. Then he turned and walked toward the barn. Chores, she guessed, but it was either too early or too late for them. Maybe he'd forgotten this morning. Odd, he looked clumsy as he walked, as if a heavy load had just landed on him. Then she felt the weight, and his hurt.

Unfair. Why did he believe in those people so much?

CHAPTER FIVE

WHEN FREYA ENTERED THE BARN a few minutes later, the place was filled with flying hay-dust. She sneezed, looked around in the dim light, and touched the noses of the horses. Then she looked outside the open back door and saw Ozzie's beautiful knee-high Space Academy boots planted in the barnyard dirt near the chicken coop. He was scooping and flinging chicken feed at the startled hens.

Her firefighting boots were made of the same costly stuff as his: man-made tekryl leather that formed to your feet and legs and could last almost forever. It was said to be as soft as fine cow-leather was, and "breathed" like it, but it was more washable. Good thing, she thought, noticing what he was standing in. He scooped more and flung it further.

His eyes were hostile when he turned to her.

"Ozzie, you can still do it, with or without Grand Galactic."

"Yeah, sure."

"Without them you are still a space trader. You believe in yourself, right?"

He stuffed the scoop into the feed sack, then stood and stared at her. He actually looked baffled. "Don't you get it? To do what I want, I need for some *other* people, besides me, to believe that I'm a space trader."

In her own mind she exploded with impatience, not at him but at them. *Any idiot can see that you already are! Anyone who can't see it is a sub-idiot! Don't wait for permission from all these sub-idiots!*

Instead she said, quietly, "I don't think you need their permission to be what you are."

In spite of the way she said it, he stepped back as if he had been shoved.

She breathed deep and added soothingly: "I will help you. We will make it happen. Tell me what I can do to help you fight this—while we're waiting for the Singer's magic to kick in."

Ozzie gazed at her. A little bit of relief lightened his face. His eyes said: You mean it. You really are offering to help.

She wanted to kiss him, make him feel better. But she thought she'd better not.

**

Norm grinned at Ozzie and Freya when they returned. He sat at the front edge of the porch, his legs dangling and bare feet swinging, and he seemed to have recovered from his own setback already. While he cheerfully polished his glasses on his almost-clean white turtleneck, Alexis was probably back at work, inside, on her greenhouse report.

"Hey listen, guys, we're on thin ice with jobs and school. And with turkeys like Raker. And even with our parents; but do you realize that when the Singer delivers the magic and the volcanoes stop, we will look good to *everyone*?

"—I mean Grand Galactic will give Ozzie back his apprenticeship, and Freya will have her job back with the 14th Reykjavik Fire Company—"

Freya snorted. "They won't need me if the volcanoes stop, Norm. But that will be *fine*—"

"—and Raker will leave us all alone, because... he's a jerk, and he'd better. Because we have the solution to the volcanoes and when people find out they'll think we're heroes—"

Ozzie made a wry face at this but Freya knew he was counting on it too, just a little. They all were.

**

At the sink Ozzie scraped the plates and set them in water to soak. *Not doing them now.* He brooded about the internship. It had been taken away from him so many times that now he was getting used to it. Only this time it was permanent? He felt like his life had just been stamped "Worthless" by someone.

Norm and Alexis had started some whispered huddle over by the porch window that was now loud enough to hear. "It just happened now," Alexis said. Ozzie glanced at them: Norm scratched his head and looked down at Alexis with a pained expression. The sun slanted in around them and turned tiny flying particles of dust into lights and stars and constellations.

"I'm sorry, Norm! My father checked into the status of my scholarship application at Oxbridge. 'They show nothing on file!' he told me. Well, of course. I didn't put in the application yet!

"So I had to tell him that I was sure it would appear in their electronic files today," Alexis wagged her head sadly.

"That means you have to..."

"Right. It's the only thing I can think of to do: immediately ask for late admission at Oxbridge under their scholarship program. If I do it now, the records will show up when I said—today."

"Pfffffff."

"My father has backed me into a *corner*." She pleaded for his understanding. "Can't delay any more. Have to apply for it."

"But you *know* that Oxbridge will instantly give you a big fat scholarship and in you go!"

"Right. I'll be riding the train from London to school in the fog every day, eight time zones away from

Berkford..." She looked miserable.

Norm grasped at a small straw: "Well, hey, we can still both start school a little late, anyway. We don't have to leave here yet—"

Ozzie sighed. "Norm, you and I have to leave *now* for the recording studio. We're late. Ready, Freya?"

**

"How come it doesn't drive itself?" Norm asked. "I never noticed that it didn't."

"It had that feature built in, but it was pretty primitive—remember, this thing was built with cheap outmoded parts, way back in 2040! So the autopilot broke long before we got it. Besides, I like driving it."

Now that his report to Berkford was finished at last, Norm was free to watch and learn at the studio today. Freya hoped he could help with some ideas. As their tires sang on the highway, Ozzie and Norm talked over the recording electronics and the possibilities for super-boosting or enhancing the sound of the Singer's voice. Sitting between them, she listened eagerly—full of questions, trying to get them to think of the next thing, and the next, fast.

"Aww," Norm complained, poking at his phone. "Message from my father... When am I arriving. He wants to help me move my robotics things to college. 'Not trying to rush you, Norm, BUT...'—Oh, I get it; moving me out of there means a room free for the little brothers.—Ozzie, remember those two thugs? When they stop living in the attic and move downstairs my dad can padlock the attic door and they'll stop climbing out on the roof at night...Jeez, 'Don't feel rushed, Norm, just get out of here.'"

Freya shrugged. "You can't blame them for making plans," she said. She could hardly contain her impatience. "About the recording?" she nudged.

In the southern outskirts of Las Cruces, something thumped against Ozzie's half-open side window: a small bird. Large insect. Bumped again. Not alive—a thing, one of those... "Ozzie!" Freya's fear burst from her mouth. She gripped Ahanith's jar tightly between her ankles.

Ozzie had seen. He stomped on the gas and turned. Freya braced against the dashboard with her hands. While the old pickup truck jumped ahead Ozzie brought his window to the top and locked his door. Norm locked his, too.

The flying thing accelerated. "Don't let it get in front of us, Ozzie," Norm said. And he muttered, "Don't know how we can lose it, here—"

Ozzie sped faster down the street. Freya knew this wasn't the usual way to PII, but he seemed to have thought of something. He made a quick left turn down a narrow street, then a right into a one-lane alleyway between two buildings, then another left that sent the deadly thing flying past them as they skidded into a spin in the little parking area enclosed by the buildings.

Freya heard the splintering screech of the flying metal hitting a concrete wall, while their momentum spun the truck 360 degrees, and twice, then three times. By the time the truck stopped, all the robot pieces had fallen from the air and the sparks were flickering out.

"Cool. Nice doughnut, Ozzie," Norm said.

They dizzily helped Ozzie gather up the pieces to present to the PII engineers. "Show and Tell," Norm grinned. Freya didn't bother to ask what that meant.

**

The three entered the PII office. Ozzie carried a large shopping bag, and Norm a smaller one.

"Look at the plants! They have grown again."

"That's right, Freya." Diana waved at the curved walls of PII's reception area, lined by stems and leaves of a generous number of potted plants. "We had to move them apart this morning after all the weekend growth, just to make room enough for each one."

Norm intoned in his preacher voice: "Excellent sign, ladies and gentlemen. The Singer is not malfunctioning."

Diana showed them into the recording studio.

"Here's what we've been using," she told Norm. The room was a small one, about 20 by 20 feet, with baffling on the walls to smooth out the sound and a velvety-looking tekryl floor. In a smaller adjacent room that they could see through a window, a large mixing desk faced the studio. In here there were a few "guest" chairs against the wall opposite the window, nearer to a standing microphone. Holoscreens hung high on the walls, nearly all the way around the room.

Ozzie lifted the Singer's jar out of the shopping bag and perched it on a stand that Freya pulled to the center of the room; the stand came up to Freya's waist, facing the mike. They waited while the engineer seated himself, got the console lit up, and prepared to start.

While Diana talked with one of the sound staff, Freya heard Ozzie say quietly to Norm, "No, I don't think I'll tell my dad about the killer cicadas. He's leaving soon. No point."

Freya pulled up a chair next to Ahanith. There was a friendly silence surrounding the Singer, as if the air around her was full of open windows and the Singer was listening through them all. [Ahanith. Tell me something? You said before: 'Parents and children have secrets from each other, always.' Why do you think they do?]

[They think they protect each other that way.]

At that moment Freya knew it was the truth. Wasn't Ozzie still trying to protect his father? And wasn't she protecting her mother from the worst of what they had learned? She asked, [Aren't they wise to do that?]

[No. Never seems to work.]

To demonstrate for Norm, the engineer played a Harriet Wane song that Ozzie liked, then a recording of the Singer's "growing song," the same song they had sent to Diana from Mars.

Diana told Norm, "At that time PII corrected the messy vibrations on the Mars recording, and had some success with it—the Singer's melody worked even better than the most effective Earth music we found; better than Beethoven. But the change to the vibrations in the earth's crust, although it was an improvement, was still small. Even amplified to the limit, the music just didn't affect the tremors in Earth's crust enough.

"Our engineers were full of hope for the recording that Ahanith would make when she arrived here. But this week of live recording and tests has not produced something that lived up to our hope—that a top-quality recording would greatly reduce the shaking of the earth's crust. And as you know, we have been recording, testing, amping, and testing over and over again. The new recordings do the job better, but not a lot better."

A screen on the entry wall projected sound in three dimensions as a holo. Diana pointed. "As you can see, Norm, compared to the usual Earth-vocal the Ahanith's sound is still complex even if we simplify it …Now here is a recording of some Egyptian music, for comparison…"

The Egyptian music was more complicated than Harriet Wane, creating a sound-graph with many more peaks and valleys, waves and wavelets of sound. "But even Egyptian music doesn't produce as fancy a graphic as the Singer's," Freya said.

There hadn't been a sound from the Singer since their arrival. "Did Ahanith forget to come?" Norm grinned at Ozzie.

[Quiet now, and listen.]

Norm jumped. The volume was low but the voice was regal and commanding. He looked at Ozzie for confirmation.

Ozzie nodded, solemnly: yes, that's really her.

Freya smiled a little. Ahanith had acted naive and almost like a newborn when Freya woke her, just weeks ago, on Mars. But here on Earth she had gone from waifish to queenly in just a week, seeming to gather dignity from the task she had to accomplish. And her strength was returning too. A good sign. Maybe it would just take more of that to—

[We will try again now,] the Singer announced. Freya repeated it aloud. The sound engineer nodded.

Freya closed her eyes as Ahanith began to sing. The music was weird to her ears, but she was long past the point where it had become wildly beautiful to her. It was wildly beautiful again now. And louder than before.

When she opened her eyes, she saw Norm giving Ozzie a doubtful look that clearly said, "This is what's supposed to stop the volcanoes?"

Ahanith stopped. [I have done it differently, to make it stronger. The strength it takes to quiet Mars winds.]

Freya thanked her and told the engineer. Across the room, Diana nodded. It seemed as if she had begun to hear Ahanith too.

They waited while the engineer checked the

recording, throwing the holo back up for them to see. Then he set it for amplified sound. "Boosted," he said to them via his microphone, and hit an intercomm. "Send this now," he instructed someone.

He brought up the seismic graphs, on 18 screens that were ranged on the wall behind her.

"This morning, 6 a.m. And 9 a.m." He showed the readings. "Seismic tremors are being recorded every hour at PII's 18 test sites all over Earth, and our graphs show that the intensity of the tremors has been going down for the last week."

He enlarged the graphs to show closeups of the last 24 hours. "Got it," the intercomm blurted. "Here goes." The graphs all changed as the last hour's results blinked on, showing a level slightly less than the hour before. And the current broadcast, of the Singer's latest recording, began to show too, as the graph angled into a slightly steeper downhill.

"Better," Norm said.

"Better," Diana echoed.

Ozzie nodded. They were all thinking, *But not enough.* Right, it wasn't.

[It's great, Ahanith,] Freya said.

[But not enough.]

[Yes, sorry, not enough yet. Again?]

"Any way you can boost it more?" Norm asked the engineer.

"If there is, we haven't thought of it yet."

Ozzie looked at Freya and raked his fingers back through his hair. He scowled, worried.

"We all will think of something," she said.

[Quiet, then. I will do it another way.] Freya repeated it for the engineer.

When he nodded, Freya closed her eyes again.

**

Ahanith had sung till she was weary. She had tried singing to the earth directly, as if it were a person. And pretending she was on Mars. She had sung the song fiercely and sweetly, roughly and smoothly and as loud as she could, always. A little louder each time, it seemed. But the graphs on the wall changed only slightly: the downhill line angled downward a tiny bit more.

"Maybe we need to be closer," Ozzie said to Norm. "What if we tried it on something close, like Oklahoma?"

"But all of these are being broadcast in about 20 places, buddy. So the sound is already about the same everywhere, close to some parts of the earth and farther from some, but on average the same."

The engineer turned his chair on its axis. "Funny, but actually there's something right about your idea," he said to Ozzie. "Look at this Oklahoma graph. We got a sensor smuggled into that area just so we could measure that volcano particularly. And look, we noticed something funny about this yesterday: here's the last week, the last 24 hours, the last hour—"

"Wow." Norm said. "Much bigger change."

"So..." Freya said, "that means the *recording* of the Singer isn't affecting the tremors as much as the live sound."

"Looks like it."

"Damn," Norm said.

CHAPTER SIX

"RIGHT," DIANA ANSWERED OZZIE, on the way out of the recording studio. "Ahanith can't be everywhere. And there's still not enough change fast enough to make the difference we need." He helped her load equipment into her air-car. "But we're still experimenting, not sure what we'll find out. PII people will keep working on this all of tonight. Some of them are begging for 'sound duty,' so this research is getting plenty of brilliance thrown at it. And we'll try again tomorrow..."

The Singer was silent in her pot again. Freya guessed she was sleeping.

"By the way, Norm, the sound engineer says your voice modulates very well. Ever thought of a vocal career?"

Norm chortled.

No one said much on the way back toward the campground from Las Cruces. Freya watched vigilantly for robot-attacks. And as usual she was adamant: The Singer should stay at the campground for safety. Ahanith should only leave Malo for recording, then quickly return again. They were jouncing along on the local road, Freya and Norm colliding as Ozzie hit the bumps without slowing, trying to get there ahead of Diana's air-car.

"Could you stop that, my friend?"

"Hey. Did you ever answer your father?"

Norm looked blank, then startled. "No."

"Think maybe he's waiting?"

"Yeah... Better let me off here at the turn. I can figure out what to say while I walk to the house, and then... call and accept my fate. There before the 15th or lose scholarship..." he muttered. "Getting seasick from your driving, anyway." He grinned at Ozzie, irritatingly.

Ozzie hit the brakes. They let the dust settle again, and Norm got out.

**

While he tore down the campground road, as familiar to him as his own hands and feet, Ozzie brooded: Norm and Alexis would be leaving before long for school, as if that was the most important thing in the world. *If today is packing day, Dad and Ilse are also leaving soon for Iceland; and that means Freya too.*

She wasn't talking about it. He looked beside him at her, wondering why. But he guessed that was pretty obvious. She had made her next plan, that was all, and at the moment he just wasn't included in it.

PII would continue to research on music and the volcanoes, even when three of them had gone. His own dreams had just been shot out from under him. Did that leave Ozzie as the messenger service for Freya's quest? Would he be stuck here, having to haul Ahanith to PII for recordings daily out of respect for an idea that seemed to be failing?

Even when the four of them were all working on the volcanoes problem together, Ozzie thought, it was clear how out of their depth they were. Smart as Alexis was, and Norm—and let's face it, he and Freya were pretty damn clever too—still they were into this way over their heads. They weren't even sure what it was that they were into.

Not that I can make it go away by knowing that it's too much. They couldn't just opt out of this part of their lives, like deciding not to go out for soccer or deciding to drop fourth-year engineering. Too late to do that, he figured.

**

As they neared the campground, he slowed his truck again. "Freya, look! How long since we've been here? Just since after lunch?" It looked like a different place. The ground under the trees on either side of the road seemed to be swarming with cats. Hundreds of cats. Maybe thousands of cats. He had never seen so many.

"Yes, visiting cats. Of course Ahanith needs rest,"

Malo said when Ozzie put the quiet pot into his arms. "She is getting stronger, though. She was very weak when you found her."

So weak she almost scared Seth off a cliff. Ozzie recalled Freya's story. *If she gets strong again, what will **that** be like? What wonders could she do?* "Do you think she will get strong soon?"

Malo shrugged. His dark eyes flickered and flamed.

Now Ahanith melted out of the pot to stand, full-size, beside Malo. Faintly Ozzie could see long cream-white leggings and a tunic, green eyes and a hint of a headdress like Nefertiti's. She was almost transparent. Like the old guests in Memphis, in the house with the ibis door.

Ozzie looked her over as a doctor would, looking for signs of strength and weakness. She was more visible, moved quicker than a few days ago. *Signs of strength.*

She walked over to the nearest large tree, just as anyone would, but then totally abandoned normal movement and rose, reclining as she did, to curl in the ample dish formed by branches dividing off the trunk. One arm dangled downward in a pose of childlike rest.

"When she's here, she likes that place, now." Malo said simply.

She seemed to be asleep again. *Sign of weakness,* Ozzie sighed.

He looked around him now. Hundreds of cats surrounded the tree, and him. All their eyes were turned steadily upward toward the Singer.

Ozzie tuned in again when he heard Freya say: "Malo, do you understand anything about the volcanoes? Why so many now?"

"Tell me what *you* have discovered," he asked.

"So much abuse of the earth's crust: explosions, drilling, rays of electricity, sound, electromagnetic waves…"

"And what do you see will help?"

"It is too late to patch the damage piece by piece. So, we need to smooth out the disharmony overall. If I was a giant… I can imagine holding the earth in both hands like a giant snowball and packing it tighter, smoothing it and settling it back down. But since I can't do that, I think sound waves can do it."

Ozzie thought it sounded a little silly right now, coming out of her mouth. Like a small girl's pretend play.

Malo only nodded. "What kind of sound?"

"Not sure. Some things help. We looked for the music of Osiris, and found the Martian Singer. We thought the Singer would help."

Was it Ozzie's imagination, or did something happen to the sound around them? Now there was no noise in the busy campground, not even the noise of an animal or a child in the background. Malo and the cats were gazing at him as if time had frozen for them all.

"It's what we think too," Malo said.

IN THE RING

**

Diana had arrived. Her vehicle sat under some trees with all doors open, like the wings of a metallic flying bug, and she moved quickly back and forth placing and setting up bits of recording equipment near the fire circle. It was the first time Ozzie had ever seen her and Malo in the same location, but she greeted Malo simply and acted as if it were an ordinary event.

The gypsies were gathering at the main fire circle, standing quietly or talking in low voices. When around 15 had arrived they sat: children as well as men and women. The Singer woke, descended from the tree, and joined them.

More arrived then, until there were more than two dozen. Ozzie had never seen so many of the campground gypsies together at once. They were mostly dark-haired and olive-skinned, like in the stories, but there were a few blondes and redheads. They all wore simple, loose clothing, tunics or cotton shirts over loose cotton trousers. Their clothes and their coloring were not unusual; those things would not mark them as gypsies if they were wandering around in Las Cruces. What would set them apart, he decided, was a kind of quiet that surrounded every one of them.

The ones seated in the circle began to sing to the notes of a guitar. The song didn't sound Martian; maybe they were warming up. After a couple of songs they waited. Ahanith began to sing, and they listened carefully. They must be hearing telepathically what she sang because there was so little volume. And they joined her after a little, the new ones learning it, singing louder and clearer, getting corrected by her so they made each sound more perfectly.

Ozzie looked for Diana. "Are you recording this?"

She nodded. "I'm doing the usual: test it live, then record, test, amp, test." She was on her phone and a headset to the studio, live-sound, for test results. "We're adding two or three singers a day to the circle, to test. It's certainly making a difference to have more people singing," she said. "Amping it helps, adding singers helps..."

**

They stopped for rest. Now Freya and Ozzie sat together behind some of the gypsy singers. Ahanith seemed to be dispirited, weak from this bout of singing. Malo went to her and spoke softly, and his words appeared to encourage her.

Ozzie thought: *Malo looks like he loves her.*

"She's getting stronger," Malo announced to all. She stood erect and returned their gazes proudly. [We will sing again soon,] she said.

Could Ahanith be some kind of long-lost "Other" to Malo?

But how can you love someone with no body? He

remembered his grandmother, so old she almost didn't have one; how he had felt for her. It seemed like love, so maybe that was an example. But Malo had stars in his eyes, like a real lover; how could you do that?

He considered further, as the singing went on: it was hard enough to imagine Dad in love with Ilse, and they were—what, 40, maybe? And here Malo was, a *very* old guy according to Diana, making eyes at someone who isn't even there! *Well, of course she's there. But not so you could put your arms around her, or*—he realized he had been shaking his head in amazement. He looked up to meet Malo's dark eyes, which were looking directly at him. His face got hot.

Malo grinned. His eyes were full of amusement.

Still, when Malo turned again to look at Ahanith, the fire in his eyes grew brighter. Ozzie struggled to grasp what he saw.

**

Their rest break was over and the circle of gypsies sang again with Ahanith. After a week of practices like this one, Ozzie could halfway hum the song himself, mimicking them.

Beside him, Freya stopped her quiet singing to point to a newsfeed on her phone. It was the Oklahoma volcano: erupting again, for the first time in months.

"Damn." Ozzie sighed and texted Norm: "Did you tell Diana what you learned about Raker and the Oklahoma oil company?"

No. Norm had forgotten. Would Ozzie tell her? he asked, and he repeated what he'd found in the Mars British colony computer files: that Raker owned a big part of an Oklahoma oil conglomerate.

Ozzie and Freya moved out of the circle to where Diana was recording and Ozzie whispered Norm's hacked data. PII was looking into possible causes of the Oklahoma volcano, she said, and the info would help. She would update them when she knew more.

**

While Diana changed settings, watched displays, and spoke into the tiny mike on her earphones, he and Freya stood by her under the trees and looked at the graphs appearing on her screen. Ahanith's effectiveness was increasing. More singers seemed to magnify her effectiveness. So did amping it up. But compared to the messy seismic stuff worldwide, the graphs made her seem like someone using a toothpick to paddle a tiny canoe into tsunami-sized waves.

The ancient song of Osiris that they had sought in Egypt, and found on Mars, and brought back at such high cost: not working?

Ozzie looked at Freya.

He saw her eyes waterfalling sparse white sparks, probably thinking just what he was thinking: *It's not going to be a straight shot to use the Singer to fix the volcano problem. Ahanith's helping, but considering how late we are in bringing her, this solution might even be horribly, miserably useless. Is there another solution?*

"We don't know," Freya murmured. "Back to that again."

CHAPTER SEVEN

WHEN FREYA AND OZZIE RETURNED, Norm was there on the front porch again, lying on his back, poking at his phone. There were the usual explosive noises.

Ozzie's own phone signaled. He pulled it out of a cargo pocket in time to see NO ID, Diana's unique way of identifying herself, disappear. A voice message instead of a live call. He played it, speeded, into his ear.

"Hey, listen to this!"

Alexis arrived at the screen door to hear. Behind her there were voices and thumpings from the rooms in the back of the house, and he could smell something cooking that was about to be pretty good.

"News already from PII, about the Oklahoma volcano. Diana says the volcano's right near the site of a fracking facility, an oil company project that pushes liquid at high pressure into rock to crack it and get at the oil in the rock. There's a big local outcry about the dangers of fracking, and the company has reduced fracking while there's an investigation.

"She says they found out a little more about what you dug up, Norm. It's what you guessed: that big oil conglomerate that Raker owns stock in *is* the one that's doing the fracking.

"*But* PII has found that the volcano doesn't seem to be affected by fracking as much as something else. Maybe fracking isn't helping the problem, but the worst vibrations are being caused by something else that's going on near there, and whatever the something-else is, *that's* what is causing the same sort of destructive vibration their instruments have been picking up elsewhere around the world."

Freya stood silently, taking all of it in. The late afternoon sunlight sifted through the leaves of the cottonwood tree, dappling the porch and Norm.

Alexis sighed through the screen. "Sorry guys. Quick, Norm, can you help me with this part?"

"She needs to finish the dang thing," Norm told them as he disappeared inside.

The old Tom walked up and stood before Ozzie at the foot of the steps. He was still scabby-looking, although he had licked the blood and the bloody hair off so he didn't look quite as bad as before. What was left was a bunch of pathetic-looking bald spots. The familiar speaking pose brought Ozzie back to cat-reality—this-is-not-a-barnyard-animal reality.

It was almost a reflex to say to the cat: [Hello. I'm Ozzie. Something to say?]

The cat's torn ear twitched. He seemed amused. Then he said, [Tom will be my alias... Ozzie.]

A cat with an alias. What next?

[Yes. And: you will see,] the cat said. [Better be ready.] He walked away with ragged dignity.

**

Freya and Ozzie had just joined the other two in the kitchen, where Alexis sat amid piles of paper and typed away on her holosheet keyboard, thumping with all ten fingers at the image of keys that lay flat on the table.

Ozzie's father walked through the kitchen fast, on his way to the porch with a couple of equipment cases. "Glad you're back! Need to talk with you, Ozzie," he called over his shoulder. "Not right now, but soon."

They were getting ready to fly. He watched Freya watch Ilse go by, pushing something that was on rollers. He hadn't seen Freya pack yet.

Ozzie looked at the newsfeeds. Always a mistake these days.

He swore quietly. The others looked. "Seth announces he's first in line for the Captain's Apprentice spot. Standings released by Grand Galactic, look: "Leading Candidate for Coveted Grand Galactic Award.""

Norm snorted and blew up some invaders on his phone.

"Oh, Ozzie." Freya looked as revolted as he felt. "That's disgusting. But now you know their true colors! If they would refuse you, and even *think* of hiring Seth, after his confessed crimes on Mars? Doesn't that tell you—"

OK, I'm lucky she cares. But in spite of that, it all came to a head for Ozzie. He'd be left sitting here with nothing: no dream, no future, friends gone, holding the Singer in a pot and wondering what had hit him—"Who gave him the right to even *be* here, to be alive on this planet and be so *evil*?" Ozzie fumed. "Let alone give him *privileges* to be that way!"

"It was ever thus," Alexis said sagely.

"What does *that* mean?" Freya stared forlornly at nothing.

"Really! Who gave him the right?" Ozzie stormed on.

"It means, *C'est la vie*, Freya. Thatsa way it goes." Norm shrugged, but he nodded sympathetically at Ozzie. "He gets the right from all the people who let him get away with it, I guess." He shrugged and punched another button to detonate the invaders on his phone.

"*I* don't agree that this is the way it should go, on and on, forever."

"No one does," Alexis said quietly. She began thumping on the holosheet again.

No one agrees. But somehow we all are letting them get away with it? Ozzie's stomach ached and his head spun just trying to grasp the problem. *Where do they get the power to do things like this?*

He felt for the ring on the chain at his throat. Hot.

There was a knock on the doorframe. He rose, but Dad strode by him with another armload and shrugged the screen door open with his shoulder. "Hiya, Marty!" he called out in surprise. "Ozzie, can you get it?"

Ozzie followed him out. Must be something they had to pay money for, or Marty would just have left it in the mailbox out at the gate, earlier today.

"Talk with you two?"

Ozzie's father grunted and set his load down on the porch. In the dusky light, Marty's face was serious. He never looked serious. "I'm not supposed to ask this, but maybe I'd better: Ozzie, you in trouble?"

Ozzie answered warily: "Hope not." And grinned like a gypsy, experimentally.

Dad turned toward him. Ozzie dropped the grin.

"OK. Well, I received a court summons to deliver to you, Ozzie. Seemed unlikely, you being a good student and so on..." He and Dad both looked at Ozzie some more, expectantly.

A summons. Holy smoke.

"Who's it from?" Dad scowled.

"From County Court, a Representative Raker..."

In spite of himself, Ozzie gave a disgusted growl.

"...Danged if I didn't forget to bring it today, Ozzie! But I'll *have* to serve it to you by tomorrow..."

"What's it for, Marty?" Ozzie's voice felt strangled.

"Well now, I don't usually read these! But I happened to, because it was so unusual—thought maybe it was mis-addressed, you know—This Raker is pressing charges against you and someone named Freya? on behalf of something or other, really way-out charges: stowaways on a U.S. interplanetary ship, says it's a *crime* under U.S. law—"

"—And this summons tells you to appear in court about it, I guess," Dad thought out loud at Ozzie.

"—It all seemed pretty strange to me, Doug," Marty offered.

Crickets chirped innocently in the wildflowers. The cicadas had stopped humming. Dad said: "Thank you, Marty." He shook the man's hand.

In the kitchen someone stirred up dinner and it sizzled. The fragrance of it had arrived on the porch. Dad and Ozzie watched Marty lumber down the steps and fade into the twilight. They watched the lights come up again inside his vehicle as he squeezed himself in. They both waved casually while he drove off.

"OK," Dad said. "We need to talk."

The screen door smacked the frame behind them.

**

Freya pushed her food around with her fork and watched Ozzie's face and his father's.

"You're a graduate and not a student any more, Ozzie," his dad was saying. "So the Academy can't protect you on this court thing. You too, Freya, I suppose: not a student now."

"Britain said it was legitimate!" Ozzie fumed. "That we were research assistants for Alexis and Norm on Mars!" Someone had put supper on the table, and they were eating, or not, as the news sank in for all six of them.

"Yeah, I don't think he can make this stick, Ozzie. But I've seen enough of these legal squabbles to know they can sure tie up your time and cost you money for a while. Don't know who in our family has the money to pay for a lawyer for you..."

Freya looked at Ilse. No one in *their* family could either. "Besides," she said, "we have to be free to—" *get the volcanoes to stop,* she almost said. She also didn't say how afraid of being in jail she was.

Ozzie looked around at the others, and he seemed to be saying it as it dawned on him: "So Raker can't win, but he can sure keep us from finding a solution to the volcanoes for a while."

Freya nodded slowly, realizing that Ozzie had just let part of the secret out.

"*The volcanoes?*" Doug said. "What's this? And why so much trouble from this Raker fella, Ozzie?"

"Oh, it's preposterous," Norm said. "Most absurd thing ever, trust us."

Silence.

"Ozzie," Freya said. "We'd better tell."

"Dad, I'm sorry. I've never seen you so happy. And I hate to spoil it. I tried to—"

"Hey. Hey." His father cut him off. "So tell me the short version. I can take it."

It still took a while for him to tell the whole story: about the damage to the Earth's crust, the volcanoes, Egypt and the Martian Singer, and the real reason for Ozzie and Freya's unscheduled trip to Mars. And how the Singer was nearby, singing daily, but somehow it wasn't enough, yet, to make the volcanoes stop.

Doug and Ilse were agog, trying hard to wrap their wits around it all. "A Martian singer," Doug muttered to himself.

Ilse said, "Can we listen to her sing?" But that was drowned in a chorus of questions, worries, resentment and proposed counter-plans.

"It's all a little way-out to me," Doug said. "I guess I'll have to mull it over to understand it better. But hey, I'm impressed with you four. Took guts to do all that. Right, Ilse?"

"Excuse me, everyone," Alexis said. "Here's something rather dumb, kind of obvious." She giggled, then went straight-faced. "Before the mail-guy comes tomorrow, Ozzie and Freya have to be out of here! So they can't be served that summons. Best for all four of us, Norm and me too, to be gone by then, so there's no one else for them to question, either."

There she was as usual, putting her finger on the

exact thing to solve next.

"That's it, Alexis," Ozzie said. "We *don't* know exactly why they're trying so hard to take us out! But we know the four of us have to get out of here—"

"Simple solution," Norm said. He leaned back and put his fork down. His was the only emptied plate. "Simple. The summons never gets delivered!"

"And Dad and Ilse had better not be here either," Ozzie concluded.

**

"How can we make this work?" Freya said. Ozzie watched her eyes go electric again, with worried pale sparks overflowing down her face.

They cleared the table quickly. Alexis and Freya tossed their holosheets down on it to make lists. By the time dusk was bringing cool air into the kitchen from outside, all six of them were ready to plan. Ilse even had coffee brewing for them all, and tea water boiling in the pot, when Ozzie stopped cold.

Dad and Ilse will need protection from accusations by Raker. He wouldn't stop at hurting them if he could get at Freya and me that way.

Ilse and Dad must not help the four of them. In fact, they needed to be in the dark about any plans Ozzie, Freya, Norm and Alexis made.

"Dad. Hey, everyone, quiet a minute! Dad." He looked at his father for a long couple of seconds. "You and Ilse don't know *what* we'll do about this, do you?"

Dad looked startled. Then wary. Then he got it. "No. We don't."

"And *we can't tell you*." Ozzie said each word carefully. He felt instantly lonely. He looked over at Freya, who nodded staunchly, backing him up. "If anyone asks, that's all you know. And here's what else you know: that we'll be OK."

An unhappy nod from Dad. Ilse gazed at her new husband, looking for reassurance.

"We'll get word to you, too. Somehow or other."

CHAPTER EIGHT

"NONE OF US BETTER TRY TO FLY from Las Cruces. Too easy to locate us if we do," Ozzie said.

Freya knew he was right. While Dad and Ilse held a huddle in the rear of the house, their hushed planning session continued at the kitchen table.

"Who has the money to fly, anyway?" Alexis spread her hands, palms up.

"Looks like *I'm* going home to California. And Berkford. Everyone will be SO happy with me."

Ozzie nodded. "Well, I can take my truck and run... Not sure where to go..."

"For me, the safest place to go is London," Alexis said.

"Still need to fly across the ocean to get there," Norm reminded her.

"Well, I'll have to figure that out next. First priority: leave here quick."

Freya stood. "And I can't go to Iceland with my

mother now. They'll find me with her and they'll make trouble for her... Take Alexis and me with you, Ozzie. For firefighting I learned to drive an old truck like yours. Wherever we go I can help you get your apprenticeship, we'll work on the volcanoes, and fight scumbag Raker if we have to. We'll get to the bottom of this."

"Let's all three go to London, then," Ozzie decided.

"Three in the truck! And I'm... going in the opposite direction." Norm's look at Alexis was forlorn. "Hey, Alexis, *you* can't drive Ozzie's truck, can you?"

"Yes, I'm sure I can. I drove my neighbor's 2024 Bentley when I was babysitting his kids."

Norm looked disappointed, but he rallied. "...Guess I'll hitchhike," he said.

"You've never hitchhiked before, have you?" Ozzie was doubtful.

"Not really. But don't worry. I will, no problemo."

They all gazed at him. Freya remembered the story of the stuffed dates in Cairo that almost got Norm's hand cut off. And then there was the "unbreakable" second lock on the Martian greenhouse. She looked at Alexis, just a little.

"Hitchhiking? Not if I can help it, Norm." Alexis giggled but she jumped into action, pulling out her phone. "Malo has to have a number, doesn't he?"

"Malo?" *Why Malo?* Freya shrugged. She thought of the exotic smells, cinnamon and cloves, coming from Malo's tent when he had left it to join them at the fire circle today. And the intricate antique carpets she had glimpsed inside. Not a very high-tech guy, by appearances.

"Why Malo?" Norm echoed. He scratched his head, leaving tufts of hair standing up straight.

Ozzie said, "I've never thought of calling Malo on a holophone. Don't think he has one. But he buys and sells things all over the state. People must have *some* way of reaching him...

"I know. My dad calls him. He uses the old living-room phone to call the campground. There's no land-line service left in our area; who knows why that still works."

**

Noises from the back of the house signaled that Dad and Ilse's private meeting was over. Ilse laughed softly.

"I can't believe your parents," Alexis whispered. "My parents would have had flying fits about this, no question. Loud ones!"

Doug and Ilse entered now with determined smiles on their faces. 'Scuse us. We want the younger generation to know our travel plans," Doug announced. "We *were* going to fly to Iceland tomorrow but we've given in to pressure from our fans"—he grinned—"and re-booked our flight so we can do a three-night farewell tour in New Mexico. The other guys in the group are crazy about this idea.

"But we'll have to take a commuter flight tonight to

be in Albuquerque in time for rehearsal tomorrow morning. We'll fly to Iceland from the last city on the tour...well, you don't need all the details."

"No," Ozzie agreed firmly.

"Good thing we're already packed," Ilse beamed at Freya, then looked into her eyes, took on a sober expression, and embraced her as if she were an equal.

Freya breathed deep. She was happy for her mother suddenly: a good man, work and music, a new life. How much confidence they gave her. Mamma seemed to have grown up.

"Airport shuttle comes in an hour," Ozzie's father said. "While we're gone, you may need to buy some...food or something." He unpocketed and handed Ozzie a folded bundle of bills. Freya watched the relief spread over Ozzie's face and shoulders. *We were just talking about this: how would we get some money.*

"But won't you—"

"No. Band earnings. We'll have enough. Now, some things to talk about, Ozzie, before we leave." He went over details they had already heard: while their parents were in Iceland, Malo would manage the campground again, and pay the bills, as he had done before. But the livestock needed to be cared for, and the garden too, at least until the end of the growing season...

I see. What he has not planned will be ours to find solutions for, Freya thought. *He is telling us that.*

Ozzie's father pulled from his pocket the multi-knife

tool and weapon combo she had seen him use a few times to cut and mend things. It was very flat and light, made of some unusual metal. Asteroid metal maybe? He handed it to Ozzie. Ozzie's eyes, now cloud-gray, got big with surprise.

Ilse reached to put a small pocket pistol into Freya's hand. "Doug gave this to me."

Freya thanked her, stunned.

"I will get another one," Ilse assured her husband with that delighted look in her eyes. "One just like it."

He nodded approval gravely.

"We are armed to the teeth," Norm declared.

**

There was time for one more cup of coffee and some quiet talk (sticking to family stories and band adventures, to avoid risky subjects) before the airport shuttle sounded its arrival notes and hung, lights blinking, in the darkness out by the gate.

Then the four adventurers played the part of good kids being left at home to take care of the place: Norm and Alexis thanked their hosts and shook hands. They all exchanged wishes for success, with school and with the tour. There were the usual rounds of hugging and tears, or maybe a little more than usual, while the driver struggled to get all the equipment and bags into the hover-vehicle.

Finally she and Ozzie helped their parents load their

guitars into the seats beside them.

They waved as the shuttle whisked the parents away to Las Cruces airport. "Wow," Alexis said, watching them glide off down the road and disappear.

"Nine p.m. We'd better haul our butts!" Norm announced. They thumped hastily up the porch stairs into the kitchen again.

"I made a list of everything we decided we must do, right?" Freya announced. "Everything on the list is marked with who does it." She slapped her holosheet onto the kitchen table. "Let's move fast, OK?"

Ozzie nodded at her and smiled with his eyes.

Her breath caught. *Saved, at the last minute, by something ridiculous: Raker. Now I can be with you.* They all leaned over the table, scanned the list and went to work. *Two can do anything,* she thought as she ran to her first task. She felt better than she had all week.

Ozzie went back outside. Into his truck he would load a good bow and arrows and some of his other old stuff from the barn to barter with the gypsies for a full tank of gas. He said they kept a large drumful for their use, always.

Freya began to gather her clothes and Ozzie's, following the tasks in her mind, searching there for anything they might be forgetting.

Ozzie would go make another deal with his gypsy friend, Qualen, to care for the garden and livestock again. But he would also tell Malo about their plan, in case they were gone a long time.

**

Norm knelt on the old brick-red earthen tiles of the kitchen floor, rolling and stuffing clothes into his backpack, muttering, "could be three days on the road: so for instant access, that's two pairs of socks and underwear on the top…" when Alexis appeared again. "Back so soon, my dear?" he said.

"Malo says someone he trusts is leaving the campground for Flagstaff, Arizona! The person can give you a lift much of your way, take you to the campground outside Phoenix. Maybe help you get a ride from there."

"Hey, brilliant, Alexis! Um, how much?"

"You kidding? Nothing. Malo says we are heroes because of bringing the Singer. You might even get free food on the way, Norm!"

Freya held up a thumb and continued to stuff clothing into backpacks. *Nice that we're heroes to **someone**.*

**

Ozzie sprinted up the porch steps, panting, and set down his gun. "OK, campground stuff done," he told Freya. She leapt up from printing things, one fist full of road maps. In a city you could GPS, but you couldn't everywhere. They had a lot of open country to go through.

"You did all that?" she pointed to the holosheet list, and they looked it over together.

"Yeah. That," he tapped the list and the item checked itself off, "That. That, and that. That too."

"Good job," she slapped hands with him and beamed.

He grinned. "You like this kind of stuff, huh, Freya?"

She admitted: "I do."

"Maybe that's why we're taking off on another wild trip right now."

"You like this stuff too! So *maybe...*"

That got a wry smile out of him.

Ozzie put on his most messed-up work jeans. While Freya stowed the maps safely in her pack, then washed up all the dishes and put them away, he would go put oil in the truck. On his way out the door he called, "Pack me some scifi to read too, will you, Freya? Anything. I've read them all anyway." Freya had assigned herself the chore of packing all their clothes, hers and Ozzie's, while he did specialized tasks.

Freya packed the pulp-book that looked best to her. On impulse she printed her long poem—the one that had won her the Icelandic poetry award—from her phone data-card using Ozzie's father's rickety printer, and packed that in Ozzie's backpack too.

**

The truck had been so cheap when Dad first bought it because it was a used 2040, and that was the last year before oil changes became outmoded. And after that, a lot of other things about gas vehicles became outmoded, too. The good part was that they could afford it. The bad part: changing the oil in the old thing was messy and time-consuming.

After a little while Freya came out to update him. He was just rolling out from under the truck. He had grease in his hair. He was a little embarrassed at his unprofessional look, but he'd done the job anyway.

"We've packed," she reported. "I'm bringing all the clothes you have except your school shirts—too easy to identify—and Alexis and I have all that we brought with us in June. She's gathering up all the food from the fridge and a polite amount of other things, like sardines, from the cupboards to take for the road—our parents will expect that, do you think?"

"Sure. We'll take them out to dinner sometime. Somewhere. And the empty fridge is good because it means Qualen can fill it with root vegetables from the garden for the gypsies to use. Part of the deal."

"Good deal I think! And now, you'd better clean up for the road. Showers may be scarce. Please come wash and dress for travel, and then, M'sieur, I will quickly cut your long locks so they are once again Las Cruces style."

She smiled at his surprise. He liked having her smile at him. She said, "You cut my hair for Mars, didn't you?"

**

Alexis was in charge of costuming. She had rummaged in the basement and assembled some odd accessories for them to wear on public roads during daylight. And stuffed them in a stained cloth shopping bag that she thought would not be missed.

Ozzie flipped through the mail as Freya clipped the hair near his ears. She was being careful not to nip him with the scissors. So far so good. When he found the large envelope from Britain that was addressed to him, he tore into it quietly. And looked, and sighed.

"Hey, Freya." He pulled out two certificates of recognition, for him and for her, from (who else) His Majesty's Government, thanking them personally for their assistance on the Mars Greenhouse Project.

Freya read them with him. "They sound so powerful. And triumphant," she said. "Too bad they are not powerful enough to protect us from having to escape from capture, or protect us from Raker's flying death-insects."

"Yeah."

She could tell Ozzie didn't want to talk about it anymore. She wouldn't say another word. She waited while he shot phone-copies of the notices, opened up her phone and his, and sent the notices off to Student Records at the Space Academy and to the 14th Reykjavik Fire Company. She knew: he did that because maybe it would help somehow. But not likely.

He entered a third address, then several more.

"You're sending to Grand Galactic?" Surprise made it come out her mouth.

"Yep. To every name I have in my phone from GG," he said. And he surprised her by grinning like a gypsy. "Just to thumb my nose at them all."

**

A pale sliver of moon hung above the trees in the west. Driving slowly, with only the fog lights on, they were nearing Norm's rendezvous spot—where Ozzie's road met the campground road. There Norm's ride would meet him soon to take him west toward California.

But by the time he gets picked up we'll be long gone, flying down the highway, Freya thought eagerly.

Alexis had ridden in the back with Norm, for time to be together a little longer. "Keep it quiet back there," Ozzie reminded them before he started the truck, but actually they had been pretty quiet for the last half-hour. *They've already said it all*, Freya guessed.

They stopped at the place where the road forked. Ozzie cut the motor. More farewells now. Large distances would soon separate them. Ozzie and she got out while Norm and Alexis climbed off the back of the truck.

Freya hugged Norm first. "Stay safe," she said. He looked startled, but he recovered fast and grinned his annoying grin. "I'll try," he said. "Stay out of trouble yourself, willya?"

Ozzie was next; then he and Freya returned to the cab to make way for Norm and Alexis to say their goodbyes.

Before she could get in, the spotted cat melted out of a nearby shadow and leaped into the truck cab.

"Where did he come from?" Ozzie tried to pull him off the seat but the old guy dug in his claws, yowled softly, spit, and finally gave a roundhouse swipe at Ozzie's arm that probably drew blood.

Freya winced at the noise. "Ozzie, Shhhh." Bad time for a fuss. "Listen."

[Do you hear me? I need to go with you!] the cat shouted, again.

Ozzie was preoccupied; he had missed the first couple of shouts. He heard, then, and warned: [Maybe not much food, maybe not safe for you...]

[Do I look worried?] the cat sneezed gustily and gave him a long stare back.

[Well, if you come with us you'd better be polite, got it? It will be crowded.]

There was no answer. The cat waited.

[Uh, you can have the place behind the seat.]

Tom topped the seat in one leap and let himself down the vertical back side of it with deft claws.

Ozzie flourished at the seat and bowed, inviting Freya and Alexis to enter.

**

They were bouncing down a dark farm road westward toward I-25, the main road, when something hit the rear window with a crack like the shot from a gun. Ozzie swerved into the other lane and stared into his rear mirror. Freya looked behind at the glass.

"Ozzie! Get your window up!" She rolled hers the last two inches too. Cracks had already spread outward from a central point on the window behind her.

Now something rapped against Ozzie's side window, traveling alongside in the airstream of his truck. The thing was smaller than Freya's fist. Its metal wings and legs gleamed in the headlight sheen reflecting from the road. And it was pushing to get ahead of them.

"Oh, no!" Alexis cried shrilly.

"Right. Damn robot thing," Freya gritted.

"Get out my gun," Ozzie said.

They did, and he hit the brake hard so they skidded to a stop. The thing soared ahead of them for about five seconds. When it boomeranged back to come at them head-on, Ozzie opened the door, aimed, and blew it up. Gleaming shards of metal exploded in the headlights and flew into darkness.

He stomped on the gas, spun the wheels and tore westward again. "Damn. I only have so many bullets. How many are out there? How do they find us?"

"Well, it's not through our phones, we know that,"

Alexis said. After a while she added, "The things have left us alone for days now. They could have—I know. They're acting as voice-activated incarceration devices, to keep us corralled. I've heard of that. Maybe a circle of them, around your land."

He nodded. "Stay quiet, then, and we'll try that idea out." he said. He arced onto I-25 now, going north, and he took the old truck up to 100 miles an hour.

CHAPTER NINE

THE HIGHWAY SLID UNDER the nose of the truck, tilting down toward them from the darkness of the eastern New Mexico hills. Roswell was ahead. There had been a fourth killer-robot right after they got onto the main highway, but nothing since then. Alexis must have been right that the robot things were there to act as a fence, to keep them near the Reed land.

Well, they were outside the fence now.

Ozzie is a good shot, Freya reflected. *Hope killer robots can't find us in other places, though.*

She said, "Here we are escaping in the dark, for the second? Fourth? time this year."

"Zillionth, seems like," Ozzie said.

"Oh, no. I still haven't got my greenhouse report done!"

Freya said, "Better write it now—who knows what's next. Ozzie knows the way, right Ozzie? I'll sleep now and drive next."

Alexis sighed, a huge sigh that seemed to include Norm, killer robotics, the journey ahead, London, and the idea of finishing the report from the bouncing seat of a pickup truck. She pulled out her phone.

**

"Why is it called a panhandle?" Freya asked Alexis. Alexis had just wakened a little while ago; she was taking her turn as driver. Ozzie slept with someone's sweater as a pillow between his head and the window while Freya balanced between the two of them, navigating.

"I guess it's because it looks sort of like one, see?" Alexis pointed at northern Texas on the paper map on Freya's lap.

"No panhandles in Iceland. You really drive fast," Freya added appreciatively.

Alexis' slow, lopsided smile turned up one side of her mouth. "No panhandles in England either, I think."

It was sunrise, but not many vehicles were in motion on the road. A hovercar passed them, going about 140 above the road surface. That might be easy for a hovercar to do, but not this old gas-truck. And not on this old road.

They had driven nonstop all night to put all the distance they could between them and anyplace where their location had been known. Now they were looking for the first of the gypsy camps that Malo had put on a list for them.

"I wish the GPSes worked here." Freya rustled through her maps and the sheaf of Malo's directions, rechecking them.

"I think we're lucky they don't. Hey! Daylight. Get out the disguises, will you?"

Good for Alexis for remembering. Freya rummaged at her feet and found the old cloth bag. She picked through it like a shopper at a flea market. On Alexis' face she installed plastic-framed glasses with no glass in them—delicately, while Alexis kept her eyes on the road. Then she found a blonde costume wig to put over Alexis' long dark hair.

"Pull over, will you? Just for a minute."

While Alexis slowed and they rolled off onto the weedy gravel, Freya pulled on an ugly black and brown knit stocking cap and stuffed all her hair up into it, then put a hat on Ozzie's head, wedged against the glass and tipped forward so the brim shaded his face. He moaned in his sleep.

As soon as the truck had stopped she looped up Alexis' fishtail braid, clipped it to the top of her head, and seated the blonde wig over it.

Then, once Alexis had them moving fast and steady down the road again, far from any other vehicle, Freya pulled down the visor mirror and did her eyes with dark eyeliner and heavy mascara from her pack. She managed it in spite of the way Alexis didn't miss all the potholes.

"Hot here," she said, itching under her cap. "The place

is outside Amarillo. There should be a sign in about ten kilometers. I hope there is shade."

Alexis' phone sounded. "Newsfeed, yours," Freya said. She pulled the phone off the dash rack to look. "Raker. Another one of his excellent news releases. Says...a ring of international conspirators is at large... those included in the ring are students...last seen blah blah. All the usual lies."

She shuddered. *Drive faster, Alexis.*

Ozzie was right: they still didn't even know exactly why Raker was suing them and trying so hard to take them out. And they didn't know why Seth had been given a chance at the apprenticeship. There had to be dozens of people more qualified than he was, even if cheaters were welcome at Grand Galactic. She sniffed out loud at the thought.

Alexis raised her eyebrows and looked sideways at her. "Haven't heard from Norm yet," she said for the third or fourth time since they had left Las Cruces.

**

The land was flat and dry-looking, and there wasn't much natural shade at the Amarillo camp but there were many tents, lean-tos and canopies set against the sun. A tremor in the earth came and went. When they found the gypsy section of the camp and stopped, a group of men, women and children appeared from among the crowded tents and surrounded the truck.

Freya pulled the hot knit cap off her curls. The men were dressed much like country men in Iceland: nondescript loose pants and shirts. The women wore leggings and tunics, belted and tied in various ways. Barefoot children, in colorful tunics over pajama-like pants, capered and peered at them.

Ozzie woke. They met the group that had come to greet them, and thanked them for the bucket of water to wash their hands and faces. Once food had been offered, the talk with their hosts was all about the trembling, which had just begun a few days ago here.

"Shaking all the way from Oklahoma to here!" Alexis was horrified.

A swarthy man with lanky arms was speaking: "We talk with the Oklahoma campgrounds, and in Oklahoma the Cheyenne, Arapaho, and Cherokee nations are protesting what the oil-drilling companies are doing. Something they are doing must be wrong."

Freya said, "I'm glad someone is taking a stand on this." And she told a little of their story, about Iceland and Egypt and Mars. None of it surprised them.

Their hosts offered them a place to sleep there that night, arrayed under a canopy where they could see the stars. But they turned it down; this was still too close to New Mexico for them to linger.

**

They plowed along over the cracked and crumbling

road, Highway 40. It must be hot as a frying pan from the day's sun. The earth tremors were steady here, increasing in force as the signs declared they were near Oklahoma City.

The shaking was weird and unsettling. But Ozzie was relieved to be anyplace at all where he could be in action again. For about the tenth time since they had left, he told himself: *OK. The apprenticeship with GG is over. What I thought was mine isn't mine.* Repetition helped. He could face it, now, without feeling like he was falling apart inside, or worthless.

But with that gone, I'm glad I'm not stuck in Las Cruces with zero future. Better to be in motion.

During the night a message had come in on Ozzie's phone from Diana:

Malo mentioned your change of plans. I won't ask where you are. Just want you to know we're here if you need us. Or PII.

That would be comforting, he guessed, if anyone could do anything to help from that far away.

"We're nearing Oklahoma City," Alexis reported as navigator. "Look Ozzie, signs for an Oklahoma Air and Spaceport."

"I've heard the spaceport's doing pretty well here, but nothing like Las Cruces. Well, we'd better avoid being seen," Ozzie said. He exited onto a road that headed south. Gusty winds buffeted the truck, which was light in the back, making it skitter sideways.

"The campground for Oklahoma City is southeast of the city a little distance," she said. "But the ground is already shaking pretty hard. The last volcano report we heard...We don't want to go too far southward—too close to the volcano. Right?"

"Yeah." Ozzie turned his gaze from the shimmering road ahead to glance at the other two. To Alexis' right Freya slept against the door, her wild hair flying out in all directions as a moist breeze blew through the windows. "Check that door to see that it's locked, will you, Alexis? Thanks. Navigate us somewhere eastward as soon as we're further from the center of the city."

Maybe for something new to do, Alexis turned on the old dashboard radio and searched around on it. She found a station that was devoted to volcano reports. Then there was another station with music and some kind of hourly news program that blasted importantly: "The search continues for four students, missing from the New Mexico area, possibly armed and dangerous. In a statement released last night, Senator B. Arnold Raker said..."

Ozzie punched the button to silence it.

Have to make sense of this. He turned what he knew over in his mind, starting from the beginning.

When they were dodging Seth's nasty pranks in Egypt, Raker only showed up to get Seth out of the hands of the Giza police. He didn't even know the four of them.

But then, between Seth and other sources, he had probably found out all there really was to know about them: students, good ones, and Ozzie a candidate for the Grand Galactic apprenticeship. *There was no dirt on us,* he reminded himself. Raker had no reason to pay attention to them.

But Raker was bothered by their attention on the volcanoes. He had not been able to talk them out of pursuing a solution. To him that seemed to be a black mark on their records. Also, when they showed his son Seth up at the robotics contest, and reported him for breaking contest rules, maybe that made Raker want revenge.

But really, revenge? So much that he had paid for Seth to follow them to Mars, and then tried to capture them? Revenge for exposing his creep son couldn't be all that was driving Raker.—And he already had his revenge: Seth would probably win the Grand Galactic apprenticeship. Raker seemed to have the power to arrange such things.

So was it the volcanoes that made him hunt them? Why would he be so determined to stop them from looking at the problem? *Even if we're wrong in the answers we find, that wouldn't hurt him. And if we're right... that would help everyone—him too.*

He stared ahead at the glare on the highway. He looked at Alexis staring numbly too. The sun on the white pavement made the dry road seem to have a sheet of water across it: a mirage, a trick of light that fooled your eyes.

He tried looking at their situation from a different direction. *Suppose Raker's actually afraid that we'll be right about the volcanoes... Like, maybe because the thing we'd be right about is something he doesn't want people to know?* He remembered Diana saying that they must be getting close to some secret of Raker's.

Alexis must have been thinking too. She said abruptly: "So he owns a fracking company. Raker I mean. But if fracking isn't the real source of the weird Oklahoma vibrations, it would help to know what *is*."

**

They stopped at the campground outside Oklahoma City at sunset. Here, the gypsies didn't occupy the entry campsites as they did at his father's Las Cruces campground. It looked like that privilege had gone to a group of veterans from the Wars, maybe the first group to become permanent residents in this old family-run place.

It was pretty flat here, too, and smelled of dried grass. The feet of a few kids chimed on the rusty swing set and climbing bars. There was an unmaintained swimming pool full of green water. And no hills or pines, like in New Mexico, but there were some tall leafy trees, dusty green, whose trunks alternately striped the nose of his truck with long evening shadows and sunlight as they jounced down the campground road.

"The shaking. It's like being at home in Iceland," Freya murmured. She stretched and shook her head to wake herself. She had been asleep for six hours, and her dark makeup was smudged but she was rosy-cheeked.

They drove among some shacks and ancient tents, and past a group of dwellings that seemed to be woven like baskets. Near the end of the winding road were tents that ringed a fire circle in familiar gypsy fashion.

Tom leaped out when they opened the door. He took off for the nearest cluster of trees at a businesslike pace.

At the fire circle they met a girl their age who was curious and amused until they gave greetings from Malo. Then she went to get the head guy for them. They were fed well: fragrant, hot fried eggs and cheese and fried bread that was spiced with something savory. Even Alexis ate all they gave her.

Grateful, Ozzie talked while he ate about where they were from and where they were going, keeping their destination vague. The head guy said his families were worried about the volcano to the south. They were making plans to move if they must.

Freya told about the volcanoes in Iceland and their search for the music of Osiris. About the Singer. As she talked, dozens more arrived to stand and listen. The teenage girl's eyes were glued on Freya.

Their hosts nodded to each other, half-smiling. He could see the light that kindled in their eyes at news of the Singer. They looked as if imagination had lit little fires there that burned through the dusky weariness on most of their faces. Taking in how thin they were, Ozzie wondered suddenly how many people had gone without supper so he and Freya and Alexis could eat.

"How are your gardens?" Alexis asked. *Good question*, he thought.

"A poor year," the leader shrugged. Alexis looked at Ozzie, her eyes asking. Ozzie nodded, and Freya had only a couple of minutes to wonder about it while Alexis left and returned—with a huge armload of carrots wrapped in damp cloth, from the bag in the back of the truck.

"We want you to have these," she said. "From Ozzie's garden."

The leader nodded approval and grinned.

The carrots caused a little delay. When some of their hosts walked with them to the truck, Tom suddenly showed up as a streak of moving spots and leaped past Alexis into the seat.

Ozzie was surprised to see new people arriving right behind Tom, moving swiftly. They wore looks that were interested but not exactly friendly. Alexis hopped in. Their gypsy host frowned and raised a blinking flashlight high into the air, turning his back to the new arrivals while he muttered: "Go quickly. There's a reward for you, and these people..."

Ozzie got it. He shut and locked the door. The man's light must have been a signal, because gypsies were already melting out from behind trees and brush to surround the truck and the hungry-eyed visitors.

Freya started the engine and let the truck roll forward without another word. With assistance from the gypsies, the uninvited visitors reluctantly got out of the way. Freya waved thanks and hit the gas sharply so they leaped ahead briskly, and without slowing they thudded recklessly over the campground ruts toward the highway.

At the highway she pushed the gas pedal harder. "Stop in a few miles when no one is around. I'll smear the license plate with dirt again," Ozzie said. "And we have a change of plans. When we ate, the guy said to avoid the Tulsa campground. You heard? There's some trouble there. He said go to the Cherokee Nation instead."

**

Once Freya had them back on the main road, Ozzie rummaged through the maps and the sketchy info on the worldweb, and between the two of them, he and Alexis got her going east and northeast toward the Cherokee lands.

"Still no word from Norm," Alexis said.

"We'll hear," Freya reassured her.

Alexis covered a worried yawn.

"Something is in from Diana, though." Ozzie waved it up on his phone again. "I told her yesterday that we felt tremors, way back in the panhandle. She says:

> **PII has been researching fast, looking for the source point of the *really* destructive Oklahoma vibrations. Pinpointed the location the vibrations are coming from! At that location there's only an obscure little company, getting 90% of its revenue from one source (never heard of the source) that is releasing vibrations into the rock layers, "testing an innovative oil-seeking technology." They've found *no oil at all* in 3 or 4 years—but they are still being paid very well by some company to go on "testing."**
>
> **And guess where they're located? Right next door to the fracking site.**

"—So we're right near them now!" Freya leaped to the next thought eagerly. "Is there something we could do about it?"

Ozzie pondered briefly. "Probably not... We're too likely to be caught, not really equipped to do much...Damn, I wish we could, though." *Better to fight than to run, always.* "Something more from Diana, just now:

> **There's probably the same kind of "testing" at other sites too, based on our global maps of the destructive vibrations. They're coming from multiple sources. We will be searching for the exact locations on each continent.**

IN THE RING

**

Freya drove onward, listening to the old radio playing stuff from local stations—a different station for each little city or town. She decided that in America people must like these. It was interesting to her to hear what music the people in these wide farmlands liked to listen to, different sounds in different places. But mostly she wanted to see if there was anyone nearby searching for her and Ozzie and Alexis.

Also on the stations, now and then someone read the news in a local voice and sometimes added interesting opinions. Just now, the name made her turn up the volume:

"California Representative B. Arnold Raker has recommended that management of the Cherokee Nation be taken over by the U.S. Government. The Cherokee Nation has filed many complaints alleging that the Oklahoma volcano is a result of irresponsible oil-drilling activities in the state. Representative Raker told reporters, 'Inability to cope with the volcano, which is after all, a natural disaster, shows that the government of the Cherokee Nation should be replaced by a more competent one.'"

Her eyes got hot, just hearing it. *Liar.*

The news-person surprised her by chuckling. "Over our dead bodies!" he said, and aired a long belly-laugh until he was cut off by a recorded advertising break.

CHAPTER TEN

THE LAND WAS RISING HERE, gaining small trees that grew in storybook bunches next to meadows of thick grass. They stopped for gas. While Alexis slept and Freya pumped it, Ozzie cleaned the windows and talked with the owner of the run-down place and with the next two customers who came in. Then he sent more road news to Diana. So far, he said, it was the first thing out of the mouths of all the Oklahoma people they had talked to: they were worried about the volcano.

Of course, Freya thought. *It's that way in Iceland too.*

**

"Good news," Alexis said later, without enthusiasm. "Oxbridge has approved my application. I'm in."

"I knew they would."

"Thanks, Freya. Yes, of course. Now I just need approval on the scholarship and I'll be ready to go. Got to have that to attend, really... Usually the notification of scholarship comes with the acceptance letter, but nothing was attached to this one...

"But where is Norm? For all we know, he could still be standing by that road to the campground, where we left him."

**

After a night of rough roads, the sun was rising. A sign said, "Arkansas 23 miles."

"Missed our turn for the Cherokee campground! Better go north from here!" Ozzie said.

There were no towns along this road so they stopped near some woods to relieve themselves and change drivers. Sporting disguises, they all slipped out on the woods side, females headed for the shelter of the tree trunks ahead.

As she returned, Freya discovered an old road sign that had fallen into thigh-high grass. Curiosity drew her over to read it hastily and take a shot of it, before she obeyed Ozzie's urging to get back in.

"It says we are at the end of the Trail of Tears," she said. "Old sign. Do you know about that, Ozzie?"

He shrugged and got the truck going. "I mostly just know space-trading history," he said. "But I did hear that a couple of hundred years ago native people were taken westward toward some land they were going to be given in Oklahoma."

Freya was already tapping at her phone: "On the worldweb, it's hard to understand," she mused. "Some of the histories are so confusing! Hard to tell whether they were slaves or members of the native tribes... But now here it says dozens of tribes settled in Oklahoma, so that must be the real story: Wyandotte, Seneca, Cherokee, Choctaw, Chicka—Chickasaw, is that the way you say it?—"

"It's the Cherokee people that he told us to go stay with, right?"

"Sign ahead says Cherokee Nation, 50 miles. Let's go. I'm hungry, aren't you?"

**

"No, our ancestors who came here were not slaves. But they were forced to come. We were driven from our homes and farms—and the Cherokee people had rich farms, with many cattle, in the valleys of the Southeast." The camp manager spoke matter-of-factly.

"Weren't your ancestors *angry*?" Freya wanted to know. She knew exactly how angry she would have been at such treatment.

They sat in the twilight at a visitors' campground among low hills, cooking some of their sliced-up carrots on the camp stove that the manager was sharing. This place was "the Cherokee accommodation for travelers in these disrupted times," the manager had said. Freya had just refilled both big water bottles from the arched tube of a pump dispenser. She rose and stirred the carrots again; they looked like a potful of steaming golden coins.

"Angry? Yes, of course at first," he answered. "Anyone would be bitter at first. And it was a hard march; many died along the way. They felt betrayed and many of the people dreaded that we might find out we were traveling this long way only to be betrayed again.

"But you must know: the Cherokee and other nations had African slaves of their own. It's hard for anyone to claim to be free from injustice to others. Slaves and slavers, freedom and tyranny... maybe the Wars were a message: there's a time to turn away from the past and do better."

A chill was settling in with the night. By flashlight, Alexis pawed through the box of canned goods they had brought, working out a menu. She looked up again and shook her head sympathetically.

He shrugged and nodded, pulling mugs of hot coffee off the lighted slab surface of the stove. Ozzie came into the circle of brightness, returned from the campground bathroom with his towel, and accepted a mugful.

The camp manager leaned back in his chair to sip at his own. "Our farms here are prosperous now. We decided to do well here in spite of the past, and we have done it. Today we have few social problems and rich harvests and those who will work for it can have plenty."

**

Early next morning, when the grass was still wet with dew, Tom had reappeared looking well-fed. He demanded entrance to the cab while Ozzie paid the campground fee.

"That's not much!" Alexis said when he told her.

"No. Half what we charge at our campground. Maybe mentioning Malo's name did that." Ozzie found a text from Diana waiting on his phone. He read, "PII is continuing to test with the Singer, showing improvement but nothing large-scale."

Freya chewed her lip.

**

Now they were on highway 40 in Arkansas, with more signs here and there that said "Trail of Tears." The bad road slowed them down a lot.

"Word from Norm!" Alexis sighed with relief. "He's... OK. Is that all he has to *say?*"

She looked disgustedly at the road disappearing under the car.

"Wait, here comes more: 'Had some adventures....Just hit Pasadena, tell you more tonight...' Oh, that beast." She tapped some keys fast.

"He *ought* to be in Pasadena, by now," Ozzie said. "Hawaii, even."

Her phone sounded. "Norm's calling!" Alexis yelped. "Norm! Didn't you get my messages? I have been so worried about you!" she began. She put it on Speaker.

"Well, I've been too busy just getting here," he was saying. "Fun, but nonstop. And then I had to unload here in town..."

"Unload what?"

"The last shipment. To help the guy out, you know."

"Oh. The guy Malo sent you with."

"No. The third one. It takes skill to unload furniture right. He said I got good at it right away. Even gave me some cash just to say thanks."

"How did it take you so long to get to California from New Mexico?" Ozzie interrupted, angling for something funny.

"Well, I had to learn to drive the truck, too—"

"Truck?" Alexis was astonished. "I didn't know you could even drive a *car*!" Alexis said.

"I couldn't. He taught me."

She gasped.

"Aren't you glad he taught Norm so you don't have to, Alexis?" Ozzie grinned. Freya began to chuckle, so Ozzie did too.

"So you were driving loads of *furniture*?"

"Only the last one. One was chickens and eggs. But see, Wally had beef and cream for Phoenix..."

It took a while for Alexis to sort out trucks, drivers, destinations, and cargoes, and by the time she did Freya and Ozzie were almost crying, they were laughing so hard.

When Norm switched off, Freya said, "Sorry, Alexis. I guess we needed that."

**

Ozzie was falling asleep. Freya could guess how he must feel, at least a little. Norm was safe in that little pink house with the flowers around it, which she remembered from Alexis' robotics holos.

And here the three of them were in a strange sort of nightmare that had been going on for days, sleeping and waking—running away from capture, on and on across a land full of worried people that was shaking from a volcano. With only gypsy campfires as islands of light.

The difference was so great it was hard to hold the two ideas in your mind at once.

Freya wished she were in Pasadena.

No, she didn't.

**

Later, when the sun had disappeared behind them and dusk was falling, Alexis announced that she was too wound up to sleep now.

She was always safe and reliable, so they let her drive on a while without a navigator while they slept like dominoes, Ozzie leaning against the window and Freya leaning against him. Ozzie dozed and dreamed and woke again to feel Freya's weight on his arm and shoulder. Actually, he liked it: kind of like being a pup in a litter. Or better. Alexis turned and caught him putting a sleepy kiss onto the top of Freya's head before he drifted off again.

IN THE RING

**

Ozzie's eyes opened. Leaning as he was against the glass, the road went by his face at an odd angle. "Still 'Arkansas Highway 40 East', Trail of Tears," Alexis murmured to herself. "Long way to Little Rock."

There was no one on the road but them, now, at 3 a.m., but Alexis wore the blonde wig anyway, and Freya and Ozzie had their disguises on too. Just in case.

He could have stayed awake, but right now all the impossibilities of this journey landed heavily on Ozzie once more. He drowsed again.

"Dammit," she groused.

"What?" Ozzie stirred and shifted upright, trying not to disturb Freya, then trying to untangle one of Freya's curls from the decorative button on the sleeve of his jacket.

"I just missed our campground," Alexis said. "But look, here's another sign for one. Maybe the same one, another entrance? Ozzie, can you check?"

While he sped up the untangling process, and reached for a paper map and Malo's list, she slowed for the second campground entrance.

They no sooner hit the gravel shoulder of the road than several armed men ran out of the darkness toward the truck. Ozzie woke fully, fast. "Alexis! Step on it. Get out of here!"

She spun the wheels, spitting gravel, and fishtailed

onto the broken pavement at the edge of the road. She stomped the mushy gas pedal to the floor. The truck labored against inertia to speed up.

Freya went bolt upright.

A bullet sang off the pavement. Another hit the back of the truck somewhere low down. Another. "Shooting at our tires! Stay down, Freya!" Ozzie ordered. *If they get the tires, we're finished.* He pictured those men, their hands on Freya and Alexis...He pulled his gun, opened his window, and shot back furiously at the knot of them.

"Someone went down, Ozzie!" Alexis' eyes were frozen onto the rear-view mirror. Their speed sagged.

Ozzie felt a little sick. Freya was groping on the floor for her pack when a bullet broke through the back window and exited again through the windshield. "Keep *moving*, Alexis!" he yelled.

Thank all the stars Alexis' head was low enough to be protected by the neck-rest. And that the shot had missed Freya.

Ozzie had never been so afraid.

More shots whined under the cab.

"Stay low, Frey!" But she turned along with him toward the back, crouching behind her neck-rest for shelter as he was doing. She put her little gun, Ilse's gift, to the bullethole in the window and shot at the diminishing figures behind them.

The shot was a little wild. "Someone else down, Ozzie!" She shot again. He saw the men hitting the ground to escape the bullets.

"You scared 'em good, anyway," he said. He grinned miserably at her.

"Maybe we both got one," she said. He thought she said that to make him feel better. But she looked like she felt awful herself.

**

They were away now. After a mile or so Alexis slowed only long enough for Ozzie to throw up onto the shoulder.

Freya looked at the floor behind the seat. Suddenly worrying about Tom, he guessed. He looked too: Tom was curled up on the floor of the cab, sleeping peacefully.

They drove on in silence, all them awake even when the morning star, Mercury, flickered and went out, overcome by gray dawn lighting the sky ahead. Finally Freya gave in to sleep, curled on the middle of the seat with her head resting first on Ozzie's shoulder, then on his legs. He watched ahead unwaveringly, giving road info to Alexis often.

He touched the ring at this throat. Still there. The smooth, heavy gold was cool now.

He had shot out of necessity. And probably killed someone. He felt like he must help Alexis now, to make up for that necessity. Right now he felt like he had turned to stone because of stubborn necessity. But every time he looked down at Freya's curls, poured across his lap, they comforted him.

When it occurred to Ozzie that the armed men might report their location to someone, he navigated them off 40 onto a parallel road. It was locally maintained, he guessed: there weren't so many potholes.

The sun rose ahead of them and pinged off the starburst of shatter-marks on the windshield. There were three more in the rear window of the cab. When could they stop to fix them? What would they do with holes in the windshield when it got colder?

Then Ozzie remembered: this truck would have to be sold before they flew to London, with holes or without. *Not clear how or where we do that.* Selling this old thing would help pay for tickets, he realized; but they'd have to find a safe place to fly from, first.

Make that soon, he hoped.

**

"Got to stop," Alexis said. It was eastern Arkansas now, and here the road was shaded by an arch of tall trees.

"My turn to drive anyway," Freya's head lifted, almost straight out of a dead sleep, and she sat upright. When she realized she had been pillowed like a child on Ozzie's lap, she turned to look at him wide-eyed, with cool blue sparks shooting at him, and smiled. He stared numbly at her. *Irresistible, no getting around it*, he realized. He wasn't sure he should be glad about that, but he was glad anyway.

At least the ground wasn't shaking here. They all climbed out and scattered to find sheltered spots. It was warm, with wet air that was so unlike New Mexico. Early-autumn sun poured down on the meadows and the tilled fields beside the road. The peace and plenty of the scene helped to wipe last night's ugliness away for Ozzie.

He and Alexis returned to the truck first. He inspected the bullet-damage: just two hits visible, and in the windows two more. Plus the damage from the killer-robot, so that made three starbursts in the rear window. He hoped it didn't hurt the sale price of the old thing much.

What if the holes draw attention? They'd better stay away from places where they could be observed closely: away from other drivers on the road, other travelers at gas stops and campgrounds...

He and Alexis sat drinking from their water bottles, rummaging through the food box for fast breakfasts, while Freya walked back toward them. She had discovered an old sign down the road nearby, and of course had to go read it. The driver-side door squealed open. "We are traveling on the Trail of Tears again," she said as she climbed back into the cab.

**

Ozzie had a new message in from Diana:

Our PII legal researcher has done some intensive public records research into the weird little Oklahoma company that's being paid to "test" the rock layers there. The "testing" company gets its money from a spinoff company formed 20 years ago by GG. Why a space-trading company would be involved in this I can't imagine.

PII will be getting help to look into this further.

When he showed it to the other two they just shook their heads. "Grand Galactic too!" Freya said. She didn't need to say "I'm not surprised." He saw it in her face.

"Grand Galactic," he pronounced the words carefully and bitterly. "Diana must be planning to get a private investigator out there..." He frowned, thinking it through: "So Raker's fracking company and the little testing company are next-door neighbors, and GG is connected with the testing company..."

"Pretty fishy," Alexis said.

Outside Memphis, Tennessee, after more delay because of detours and road repair—Ozzie told them, "Let's just go around this and keep moving," so they kept taking side routes—they finally found the city campground east of downtown.

It wasn't on Malo's list, but a waitress behind the counter in a diner had suggested it when he stopped there to ask a couple of hours ago. She drew them a map, and took the time to ask around the restaurant for names of roads and places along the way; Ozzie was so grateful for her plump, reassuring face that he brought the other two in and they sat at her counter. He splurged on burgers and hot apple pie for them all for supper.

He was glad again for that reassurance when they arrived at the Memphis Blues Campground long after nightfall and rolled quietly up to the dark office at the entrance. Especially when floodlights suddenly turned on and lit their front seat and their faces. Good thing they had the disguises on.

A bulky figure in overalls plodded through the office door toward them, sleepily. The floodlights dimmed gradually, leaving a little bulb glowing at the door.

"A place to stay?" Ozzie asked. "Just the truck here. We have friends in the gypsy camp."

The woman drew a small map on a pulp-pad, tore it off for him, and nodded for Ozzie to drive on.

Things looked friendlier as they neared the X on the manager's map: a crackling fire spat sparks in the center of a clearing, and when they pulled up close and parked, Ozzie could see a ring of men and women who looked familiar, a lot like the gypsies back home, with their faces glowing in the firelight. He opened the window a little and heard laughter and a pleasant murmur of voices.

**

Freya, Alexis, and Ozzie stood in silence looking into the eyes of the head gypsy, a woman whose age she couldn't guess. This silent looking was something Ozzie had taught them after the first gypsy camp. Maybe, like Malo, the woman was ancient? Freya shut off the thought. She might hear.

The fire circle fattened to three deep as they ate and Freya told the story about the volcanoes, Egypt, Mars, and the Singer. "Now we are being chased for no crime at all," Freya concluded. "So unjust." She felt full of injury and bitterness right now—for Ozzie and herself and all people who were treated wrongly.

There was silence. Not unkind; it was a respectful silence, really. Alexis looked around at them all and said, "We have been driving along the Trail of Tears. You know about that?"

Some of them nodded.

"I wonder: were the gypsies ever mistreated?"

The head woman said, "We have been driven out of many parts of the world. And told to leave others. And sometimes shot at, stoned, or attacked some other way."

"Terrible," Alexis said. "Why?"

"We never knew, but we guessed. We are different."

"How hideous. And were you captured, or enslaved?"

"Captured sometimes. Never slaves."

Alexis was surprised. "How have you managed that?"

"We would kill first. Or die." She smiled a fierce and joyful smile.

"So you never allow people to mistreat you?" Freya asked wonderingly.

The woman searched her face for a few seconds. "Do those ants walking over your boots mistreat you?"

Freya looked to see a determined line of them marching across her boots. How could the woman see ants from all the way over there? She looked around at the firelit faces, grinning at her. Were they mocking her? But they didn't seem to be.

The woman said, "Does a cold wind mistreat you? Or a scratching cat?"

Freya looked at Ozzie. She was baffled. A wind or a cat was not the terrific sting of human cruelty. But he was no help. He just gazed at the woman. He seemed to be trying to understand.

The woman persisted: "If your horse tries to throw you off, do you let it mistreat you?"

Freya started to get it. "No. I would stop it." She spoke from her recent experience.

"So. We do too. When we need to, we stop it."

A few of the other seated people nodded. Someone began to strum a guitar slowly, gently. And then someone began to sing.

The song rose, swelled by more and more voices. The harmony was eerie and beautiful. They sang without holding back: it was the mournful, voluptuous singing of people who know happiness and sorrow and know that really, the song is everything.

Tears ran down Freya's face, although she didn't understand why. Ozzie and Alexis were looking at her now, so embarrassing. She put her head down on her knees and cried as quietly as she could so she didn't spoil the singing.

CHAPTER ELEVEN

IN THE PASSENGER SEAT, Freya woke to find Tom on her lap. It startled her; when did this one get so friendly?

[Don't flatter yourself. Just helping out.]

The road noise was a steady thundering beneath them. They were driving through thick woods, near Nashville, the signs said, with thicker, greener growth than she had ever seen. The road was like a slice cut through the tall greenery. Then she remembered her dream:

She wore an armored tunic. A helmet was heavy on her head. She saw the enemy approaching, and it was a wave of molten rock rolling forward. It made the noise of chaos. It would not stop. She lifted the sword high and it began to sing music across the molten wave, dividing the wave in two.

She didn't want to but she walked through the slash between the waves, with towering red lava on either side, and the sword sang as she went; the music cut deeper so she could walk onward.

The noise and the song of the sword now dissolved as she walked. The wave and the tunic and even the sword disappeared.

A planet turned in the silence of space, floodlit by the sun. The planet hung in a net of lights, a big fishing net with a lamp attached at each knot where the strands met. And then as she neared she heard the net sing: a brilliant, beautiful sound.

She could almost hear the music now, but not quite.

Tom strolled off her lap and leaped over the seat to his place behind it.

**

The hand-lettered sign said "Home Grown Fruit and Vegs" and a list of offerings was propped up beside it: apples, squash, lettuce, beans. Alexis had stopped without consulting the others, and probably just because she was drawn to the fresh-grown things, but Freya knew Ozzie was as glad as she was to get out and see what was there.

Now Alexis was bagging up a few handfuls of string beans. "We can eat these raw as we go," she explained. Freya nodded and held up two hands full of gold-speckled red apples.

"Hey, you the kids with the greenhouse?" The man

tending the stand looked back and forth from Alexis to Freya. His eyes settled on Alexis. "You did that greenhouse on Mars, right?"

Freya saw Ozzie freeze. Alexis nodded.

"Sure thing! So it's really you. Saw the pictures in the ag mag. Agriculture, yeah, farm magazine. That was pretty clever, a greenhouse on Mars. For you, just fifty cents."

"Raker's nasty publicity has made us famous," Alexis muttered to Ozzie as they closed the truck doors. "But damn. Should have worn the disguises."

"Fast, Alexis, so the guy can't figure out the plate number..."

Alexis got the truck engine started fast and hastily turned back onto the two-lane road, did a polite job of peeling out, and made it up to 80 quick. Throughout all that, Freya watched anxiously in the rear-view mirror.

Still, she was thinking: Raker had made them famous, even got them free food. What if Raker somehow ended up helping them?

What a curious idea. It seemed like a familiar one but she couldn't quite remember where she had heard it.

**

It seemed to Freya that the further they traveled, the more there were radio announcements about searches for the ring of conspirators. Right now while she drove she tried to count how many she had heard.

She gave up and touched the search buttons, looking for a nearby station.

"Why are you always listening to that radio?" Alexis sounded irritable.

"To see if they're looking for us in each place." Freya added testily, "Why are you always checking on your phone?"

"Looking at ways to get to London if we have the chance: little airports, you know, between here and the Atlantic, that maybe we can escape from."

"Good idea. Maybe if they get tired and stop chasing us for long enough, we can fly from one of those. Of course, we'll have to get the money to pay for tickets first, somehow..."

"And if we don't, maybe I'll find some other way to get across the ocean."

**

"More from Norm, Freya!" Alexis giggled. "Listen to this..."

Freya interrupted. "Alexis, did you get the scholarship yet?"

Ozzie heard the question as he was drifting off.

"Not yet." Alexis seemed oddly unconcerned.

Almost gone, Ozzie heard Freya say, "Got a message from Mamma and Ozzie's father. Landed in Reykjavik. Whoo. I'm glad they're safe."

Reykjavik safe? What is she thinking?

Safer than we are. He submerged, with that thought anchoring him downward, into the tide of dreams.

...He was surrounded by the beauties of the place, the music everywhere, the iron-reddened soil, the thick green leaves and feathered ferns, the grace of winged lizards and the songs of the small lighted dragon insects that put magic into the nights.

It was night now, and the air was sweet with blossoms and full of insect songs, the boom and hiss of the flying creatures, the sound of water in the irrigation ditches. Like distant singing, there was a throbbing beneath the sounds that was woven through all of them...

He moved away, so far away that the lights and sounds became a net that clung to the turning sphere of the planet. The net pulsed and blurred, and the music was everywhere.

Then he moved inward, closer again, and the lights again became clear. The blue seas of Earth shimmered, caught there in the sparkling net of sound.

**

He drove now, with Freya as navigator. The road cut through hills and thick forests here, and many little towns and cities. It seemed safer driving after dark like this.

Now that they were in more populated areas she had given up on the radio stations because they played the same music in one town as another, she said. She checked her phone for newsfeeds instead, calling off the news to him now and then. Being her own radio station.

They were about 5 and a half hours into his 6-hour turn, and he would be glad for a rest but right now he was driving at top speed, dodging the bad spots, seeing if he could get them to the Nashville campground by the end of his turn. They could all use some horizontal time again.

"There's not a word on the newsfeeds about the volcano in Oklahoma," Freya mused. "Wait." She gasped. "Lava leaks in Japan, at Mt. Fuji, Ozzie! I hoped we would never have to hear that..."

He bit his lip.

"Raker newsfeed in too—"

Ozzie waited, wishing not to hear it.

"Oh... It says all airports alerted, 'international ring of conspirators to be prevented from flying'..."

His empty stomach turned to lead. "*All* airports?" He said numbly. *Trapped.*

"Good that we're driving, and we didn't try to fly—"

"—Yeah, but how are we going to get to London if the airports are closed to us? Row a boat?"

"How about a private hire, my friends?" Alexis was awake now.

"You're supposed to sleep, Alexis."

She had to know that she needed more sleep. But she ignored him. She shook out her hair and brushed it and rebraided it. *Funny how some women do that. When things hit a roadblock, they redo their hair.* Something he'd noticed. *Maybe it gives them time to think.*

Now Freya also pulled a hairbrush from the pack at her feet—her footstool as they drove—and brushed her unruly curls back into a new bun at the nape of her neck. While she tugged and smoothed, she gazed unseeing into the apron of road that was lit by their headlights. Her eyes made a soft waterfall of pale green sparks in the dark cab. *Proof of concept: she's thinking.*

He was thinking too: of boats, of smuggling themselves via Bermuda and the Canary Islands; of spacecraft...

"How could we possibly afford a private plane, Alexis?" he said when she began to tap at her phone.

No answer. She went on tapping quickly while dawn grayed the sky ahead of them. Freya pulled the gum out of her mouth and stopped up the hole in the center of the windshield. The inrushing air was cold here.

**

When they were too late to sleep at Nashville, Ozzie went on driving overtime to let the flight research go on unstopped for a while, then made Alexis take over so he could sleep. The roads were rough here, whatever this place was; most people who passed them were in aircars or other airborne vehicles. *Maybe that's why they don't bother to patch the pavements.*

He wakened briefly again east of Louisville, Kentucky, his head mashed against the cold window. "The sun is blinding me, Alexis!" he heard Freya complain. He opened his eyes a crack. He saw the morning sun high in the sky and a smear of green farmland sliding by. At least the gum was still there blocking the airhole. Alexis didn't answer; she was on her phone, tapping again. Or still.

**

"We're here, Alexis," Freya repeated. Ozzie drifted awake again, as she slowed, and saw that now she had turned in at a large grassy meadow by a sign that said Marianna Campground. It was one that Malo had put on their list. "InDiana, Ozzie." Alexis had dozed off with her phone in her hands.

"That's 'Indiana,' Freya." They drove down a gravel road between lush green fields, for two miles or so. There the gypsy camp, obvious because of the arrangement of tents around the central fire, was surrounded by a wreath of rich fields and gardens. Ozzie sat up belatedly and said, "We don't have to ask about how their gardens are doing *here*, do we."

Maybe we can stay with these people for a day or two and sleep lying down for a change, while we decide where to escape to next, on our way to London. Canada, maybe.

A group had gathered to greet them, standing or sitting at the fire circle, some of their faces shining with sweat from the morning's work. Freya told them about the volcanoes in Iceland and Oklahoma.

They had known for some time. And they knew about the Singer already. Their communications systems were surprisingly good. Ozzie half-listened, still thinking about airports and transportation.

Their hosts feasted them on thick stew and eggs, all rich and filling. Ozzie approved of every bite. While they ate the spokesman looked all three of them over, as if he were weighing them. "It would be better for you to rest I think," he said. "But there were investigators in the nearest town just yesterday, looking for 'young conspirators.' This was how we knew you must be coming soon. Best for you to keep going fast."

Freya looked at Ozzie. Her eyes said wearily, How? And where to go?

One of the gypsy women disappeared briefly and returned to them with a dark, bluish-green gem in her hand. *An emerald?* "For your travel," she explained. She polished it with a soft fold of her tunic, murmuring quietly to Freya about where to sell it in New York City for cash.

"But that must be so valuable!" Alexis protested to the others. "Don't you need..."

The woman shook her head. She said, "Malo sent a message that you would require this. There's a small cave in a location no one else knows about; we take gems out of the rocks there. They cost us nothing. Pretty gifts from the earth..." she dismissed their value with a wave of her hand. "I sell some, in one city or another, every six months, and we keep the source a secret."

Still Ozzie was embarrassed by such an expensive present. True, the three of them needed it, but they would have to arrange to repay. Before he could speak she turned to him and said, "Malo told us what you have done. You may not know yet what all of it means. This is *small* pay from us." Her hands swept outward to include the others around her.

Ozzie was relieved by this new thought: To Malo and these people, they had earned an emerald. *All our crazy, maybe useless struggle to get the Singer to Earth: it seems to be worth a lot to them.* He watched Freya's eyes as she wrote the directions down carefully.

Almost before they could think, they were back in the truck and in motion again. A few miles later, Alexis said to Ozzie, "About what that woman with the emerald said: It's kind of odd that they care so much.—Not wrong, just odd."

Ozzie considered, then said, "Yeah. They're not just being kind to us. The gypsies seem to have something of their own that's riding on the Singer, and on us."

Silence for a while. "Well," Alexis declared, "we still need to find out how to cross the Atlantic without an airport." So Ozzie drove, Freya slept; Freya drove, Ozzie slept. Past large flat farms and along nameless rivers, Alexis insisted on navigating while she prodded the worldweb for a solution.

**

Ozzie woke. It must be almost noon; the sun was high. He had heard something interesting. "Where are we?" He sat up quickly.

"Yesss. Getting somewhere..." Alexis said. She pulled out one of her puzzle books and tore the blank pages from the back. "Knowing that we can't afford any *normal* private flight, I've been looking for some other kind of vehicle... Like this one: air freight, one of those air trucks that carry deliveries..."

"Really?" Freya said. She yawned and wriggled her shoulders to ease them, then gripped the wheel tightly again.

"Freya, pull over. My turn. You need to sleep, Alexis," Ozzie said. He looked over the back of the seat to make sure they hadn't left Tom somewhere. Although it might be more fun if they did.

"So this says he's revolutionizing transoceanic delivery..."

"Like delivery to Europe."

"Or Britain. So he says—"

"Delivery of what? Maybe he doesn't do people," Freya said.

"Yeah, just a minute, Freya. He does people, yes. He

says he's doing this by boosting an ordinary airtruck with older, slower space drive technology so it goes faster than with jet fuel, but not as fast as space travel today, like maybe not as fast as GG goes to the moon, right Ozzie? And he just uses it for about five minutes, to uh, leap across the Atlantic—"

"Whaaat? Sounds hokey. Never heard of that." Ozzie scowled and stretched. "Freya, pull over, time for me to drive."

"And he says his prices are low because this is a new approach and he's still getting started—"

"Does he give any references?" Freya wondered. "You know, someone else's opinion? Five-star rating or anything?"

"Hmm. He has 5 stars, but that's from only two reviews."

"Could be two of his cousins." Freya stopped on the shoulder and craned her neck curiously, to see the image on the phone.

CHAPTER TWELVE

"STOP PRETTY SOON, OZZIE, OK?" I'm ready to pass out," Freya said. She was shifting around in the crowded front seat, arranging her things so she could take her turn sleeping.

"Why is this called Ohio?" Alexis muttered. She talked without stopping, the way people do when they're hectically tired: "Freya, I still haven't heard from Oxbridge about the scholarship, so my parents text me every single day to ask about it. If they're worried about *that*, think how they'd feel if they knew this is the way I'm traveling to London? Crazy.—Hey, newsfeed from Las Cruces. Want to hear this, people?" Alexis didn't wait for an answer.

> **The Humane Society of Las Cruces, New Mexico has reported a large influx of cats into the area around the city, which has earned the nickname "Space City, USA." There the local dog population**

> remains relatively stable, but cats seem to be off the charts.
>
> The spaceport itself, Spaceport USA, reports the presence of many cats who are showing up at points of embarkation for Moon and Mars flights. Said one of the runway crew, "There are some good mousers here, leaving us 'presents' to show us they are at work. As a testimony to their success: there are fewer field mice showing up in the traps of the spaceport restaurant kitchens."
>
> One or two of the cats have accidentally wandered onto the ships and become unwilling guests of a Mars Sciences flight—

"—I can't believe that!" Ozzie shook his head.

"*I* can!" Freya slapped the ancient dashboard with delight, bouncing a cloud of road-dust into the air in the cab. She felt much more awake now.

> —but the spaceport has assured the SPCA that they were sedated for their safety as soon as they were discovered aboard the Mars-bound ship, and on arrival they were delivered to various colony buildings on the Red Planet. There they will make new homes for themselves and help balance the ecosystem of the mouse-infested Mars colony. There are already a few cats mousing in the Mars Colony...

Another spaceport hotel employee said, "I've seen so many cats here during the last month. They're very polite, and they never seem to get into the rooms but they sure do catch mice.

Alexis thumbed to scroll the screen upward.

"It says, cats have been taken on as mousers in most hotels in the state...

"Here's another story—a spaceport public relations guy says:

We always hope that visitors to the state will decide to visit the Spaceport as well—in every hotel we put a little holo display about Spaceport USA. This month, hotel guests signed for spaceport tours in record numbers: particularly for the Mars colony preview tour. And in spite of the cost, Mars ticket pre-sales have doubled."

Tom leaped from the top of the seat-back onto Freya's lap. He sharpened his claws on one knee of her jeans, gazing at her intently with mouth half-open and his ragged ear drooping. She listened, but he said nothing that she could hear. She decided that he was grinning at her.

**

Freya had her head against the window, with someone's rolled-up sweater against the cold glass. It was growing colder as they drove northward; right now she wore her own two sweaters, one over the other. She was already drifting toward sleep when Alexis' voice came to her from far away.

"Norm's at Berkford!" Alexis announced. "Got a weird roommate...food is OK... seeing his advisor tomorrow." She giggled at something she didn't read out loud, then added soberly, "Well I'm glad he's safe, anyway."

**

The dark road fled under them. Freya said to Ozzie, "An answer came this morning from the fire company. The chief got the commendation you copied to him, about Mars."

"Yeah? What did he say?"

"He said 'This is what you did about the volcanoes??'"

She thumped the steering wheel restlessly in time to the music Ozzie had going.

"It does sound a little outrageous, Freya," Ozzie admitted. Alexis giggled sleepily.

Freya nodded and hit the gas harder.

**

It was quiet in the cab. Freya had just finished telling Ozzie the story of Yggdrasill, the giant tree at whose feet the gods met, and of the Valkyries, who carried fallen heroes home to Valhalla to feast and fight for fun without suffering.

Alexis had finally agreed to sleep. Ozzie, as navigator, sat with the maps on his lap, just in case; they would be in Pennsylvania soon, and alerts about them were on too many radio stations. Missing a turn could be serious at this point.

"Thanks for bringing us," Freya said.

"Huh?"

"Away from Raker and his legal papers and his death-insects..."

He snorted. "Well, I was coming this way anyway, right?" But he looked at her, wondering what else she was thinking. The road was rough; the wheel vibrated in her hands. She was holding the truck in line, a hard trick, holding it this steady so she could do—what was it?—85? on this piece-of-shit road.

After a few minutes of silence she let him know what was on her mind. "As soon as we can breathe again," she said, "we need to work on your space-trader plan. Right? I mean, don't you think we should?"

He looked again, saw her hair flying in the small breeze from the cracked window and her intense, eager eyes. His heart melted for her. As boundlessly tough as she was, something about her seemed fragile too, at that moment. How could she be both?

He didn't want to move. "Yes, we should," he said, and that was all. He saw the lights come up in her face and body as if someone had adjusted a control knob.

So he added, "Thanks, Freya."

In some parallel world she leapt into his arms. He felt her leap, like a dancer, for him to catch her. If her hands were not glued to that wheel, her hands and arms would be around his neck. He closed his eyes and in his mind he caught her.

When he opened his eyes again, she sat there as before, watching the road, still gripping the wheel tightly. Her eyes were the giveaway, though. They shot blue sparks for yards, out over the hood of the truck and into the night.

**

"What are you reading?" Freya asked Ozzie. He was slowly turning the pages of a fat sheaf of papers lit by a little clip-on glowlight.

"Your poem," he said. "I found it in my pack. You must have put it there? You didn't say that you translated it all into English. Makes me want to go to Iceland."

She looked at him for a whole second, then back at the stark fan of their headlights spreading across the dark road. She felt like someone swimming in moonlit water. Happiness washed over her.

"Thank you, Ozzie," she said.

CHAPTER THIRTEEN

OZZIE SHOWED HER: they had almost none of the money left from Doug and Ilse's roll of bills. But there was some food left, and the truck. It was Freya's idea to stop in a town park somewhere outside the famous city of New York, which they couldn't even see from here, before they got to where the huge buildings and crowded streets were.

They sat in the park, inside the sunny truck cab with the sun-shades down to screen them from snoopers, and ate heartily: the last sardines and tuna and the carrots. And tekryl cans of soup that they passed around and spooned out, cold. Bits of light things that were left, like protein snacks and dried fruit, they stuffed in their packs.

Somewhere in Pennsylvania they had chosen Alexis for the next part of the plan. In a gas station bathroom (one quarter tank only, Ozzie directed) Freya now made Alexis up to look older and tougher. Ozzie coached her on the rules of negotiation. As usual she was a fast learner. Good thing, Freya thought; the truck had never been pretty, and it was worse with bullet-holes.

They drove to an area in the city where the worldweb listed many Asian names and businesses and asked around to find the most popular car dealer. They hoped the dealer might like Alexis because she looked familiar, and might want to give her a break because she could speak Chinese. Ozzie removed the license plate from the truck; he and Freya took all the backpacks and tried to look fairly invisible in a nearby fast-food place while Alexis did her best for the next couple of hours. She got a few hundred for Ozzie's wheels.

When she entered and found them in a booth, they handed her sandwich over. "You got as much as I would've, Alexis," Ozzie said. "Don't worry."

Tom was in a box that they had folded and punched with hand-holds to make a carrier. Freya had been studying train and bus maps. "Here's where we're going," she said. They helped each other put on the packs, walked to a subway entrance, and got on.

At the Harbor Hostel, they paid a little for showers and a night's rest. Maybe more nights would be needed but Freya hoped not; she saw the bugs that ran when she turned on the bathroom light. They all cleaned up totally, including their nails. Everything clean. Freya lent Ozzie her shampoo.

Then Freya went in disguise—nice clothes, hair up

under a hat, and heavy makeup to cover the freckles—to the jeweler who would buy the emerald. Ozzie insisted on having Alexis waiting safely at their room in the hostel with the city maps so she could arrange their flight and help from there if she needed to. And he insisted on escorting Freya.

But he did it by following her, on the opposite side of the street, about half a block behind. She wore a bright scarf for visibility and her best jeans and walked fast, down twelve blocks of sidewalk toward the jeweler.

She noticed old-fashioned cameras on some of the light and signal poles. Some looked like they could still be working; others had shattered lenses, as if someone had shot them out, and some dangled uselessly from their wires. She had read about these cameras, in use in America and Europe decades ago. They made her uneasy and happy to slip into the store.

**

Ozzie and Alexis counted up the money. "He wants $600 for three of us..." Alexis told Ozzie.

Freya tossed up a holo from her phone. "Oh, Ozzie, look! Let's go see her! Want to, Alexis? The Statue of Liberty, in New York Harbor."

"Freya, you got $400 dollars for that emerald?!"

"Shhh, Alexis."

"Sorry, Ozzie, but who will hear? That's a lot, Freya."

"Mmmm. Do we have enough to go?"

"Yes! We can fly with the delivery guy."

"Good—Good, I thought we could. I mean enough to go see the Liberty Statue. Do we?"

**

"Not going to leave any of it here," Ozzie said. They made a money belt of Freya's bright Egyptian scarf and wrapped it lengthwise around the stack of money, then tied it snugly around Ozzie's waist. He complained that they were tickling him.

Their bus left them near the address the overseas delivery service had given to Alexis. "It looks like just a house," Freya said when they arrived.

"His science seemed OK, though, didn't it," Ozzie said.

"Brilliant, actually." Alexis gazed up at the face of the old house. "Said he's done this fifty times."

The person who answered the door looked like he was about their age.

"We'd like Overseas Delivery Agency," Alexis said in her most businesslike voice.

"This is it," he said. Freya's hope sagged a little. "I'm Thoren Bisk, President. You Miz Wu?" He stood looking at Alexis, his face as disappointed as Freya felt. *But we don't want to assume that young people are idiots,* she thought. *I'm not. Maybe he's not too.*

"How soon could you fly us to Heathrow?" Alexis asked.

"Not Heathrow, at the price you asked for," he said.

"But nearby. Walking distance. Show you how to walk in when we get there. Are you prepared to pay now?" He asked. "I'll need to stock fuel…"

Alexis looked at Ozzie. Freya asked, "What about half now in cash and half before departure, tomorrow, also cash?" she said.

He considered for a couple of seconds. "That's fair," he said. He had just passed her test: smart enough to know right away what fair is, and that this was the safest method of payment for both sides.

"OK with me," Ozzie said. He smiled at her.

"OK." Alexis repeated: "Tomorrow."

Thoren had a car, he said. His machine was in its hangar at a local airport. He would take them there tomorrow if they would attach their own packs on top and pay for some gas for the vehicle.

Ozzie shrugged at the other two. They all shrugged. Sure, might as well. "Leave here early tomorrow, right?" Freya said. They went inside to get the money out of Ozzie's money-scarf.

"Are we crazy?" Alexis said when they were on the bus again. "I know it was my idea. But. All that money for a hovercraft that isn't really designed for transoceanic flight."

CHAPTER FOURTEEN

THEY TOOK A BUS TO BATTERY PARK and there she was, standing far out in the harbor. Freya gazed at the statue a long time.

She saw that Ozzie had brought his truck license plate wrapped in a fast-food bag; when he thought no one was watching he stuffed it deep into a trash container at the water's edge.

Then they walked and talked along the waterfront, read the legends on signs, ran a couple of races, lazed on the grass, and took pictures. They did some of those joke holos that perched the Lady on Ozzie's head and on Alexis's hand. They did one more with the boosted telephoto on Alexis' phone, of Freya holding her water bottle up, looking the same size as the statue that raised the torch in her hand.

Ozzie showed her, on the bus. Maybe years later, Freya thought, they would look at the holos they had just taken, and Freya would see herself again as he had captured her just now: they were on their way to meet disaster but for the moment she posed fearlessly with her head held high, face-to-face with the Lady of Liberty.

**

When they first saw the Overseas Delivery Agency vehicle up close, Freya had to suppress a gasp. It had looked much better on the worldweb. It was really just an odd-looking air truck with some unidentifiable attachments on the back. And were those wing-things really going to stay on?

However, they had taken off. So smoothly it was surprising—to all of them but Tom, in the makeshift cat-carrier, who was already cross at the delay about the Statue of Liberty yesterday, and at being left in the hostel room alone for hours with nothing to hunt but bugs and very ill-fed mice. When the flight went into its high-speed leap he lost his temper and yowled furiously at them all.

Now Freya watched Ozzie's eyes, fixed on the ocean through his window at the left, and she realized that it wasn't worry that had riveted his attention; it was just that he had never seen the sea before. She looked past his shoulder, enjoying it through his eyes, and tried not to watch their pilot too much.

But Alexis, having decided that they had a good chance of making it across the Atlantic safely, was busily tapping at her phone, tossing icons into the air as she sorted, rapid-fire, through her mail:

"Hey, another newsfeed on the Las Cruces cats. Want to hear?"

Freya nodded gamely.

> **More news from Las Cruces, New Mexico, where cats are still a local phenomenon:**
>
> **A terminally ill six-year-old boy has made a dramatic recovery after his family took in a stray cat to keep him company. He claims the cat made him get better and that now he intends to "grow up and fly to Mars."**
>
> **To assist him, Grand Galactic, the world's first and largest space line, has donated a ticket to Mars for him and a friend—and the cat.**

"Great publicity for Grand Galactic," Ozzie said bitterly.

"And here's another one. About our friend the cook!"

> **When a supplies storage capsule arrived at the British Mars Colony kitchen, the cook opened it as usual and was astonished to find the 60-gallon container filled with cats.**

"Oh, that's perfect!" Freya crowed.

How the stowaways managed to hide themselves in the container is a mystery. Two of the furry travelers didn't make it to Mars alive, probably due to the extreme cold in storage enroute. But one who had been sleeping beneath them gave birth to six kittens shortly after arrival.

"We have room for them all," the British Colony cook said. Her kitchen is the first on the colony to grow all its own vegetables, and the British staff on Mars welcomed the arrivals, because they seek a more mouse-free kitchen, greenhouse, and quarters...

Freya heard between the lines: two cats had died to protect the pregnant one. A very practical solution, of course. The cats would have embraced their duty with irony and they would have just... done it. How like them to do that. *Wonder if I could be that brave.*

Alexis snickered, then rocked her seat badly as she shifted around again to turn toward Freya. The vehicle rocked too.

"Hold still, Alexis, willya?" Ozzie groused. "This thing is wobbly enough already." Their pilot scowled. Their packs were stuffed along the sides and behind them, without enough room to make it a balanced load.

"Sorreee. Listen: 'Chief animal behaviorist of San Philmo Zoo says cat migration to Mars is a natural result of the recent cat overpopulation in the Las Cruces area.'

Scientific explanation." She giggled.

"Lightheaded. Need an oxygen mask for her?" the pilot said.

Somewhere at the bottom of Alexis' message stack there was one from Oxbridge: "Finally: Oxbridge digs deep and gives me a scholarship. Just in time to make Mother and Father happier to see me." But she didn't sound happy herself.

Freya said so.

"Right..." Alexis admitted. "The longer the scholarship was delayed, the more I hoped that I wouldn't be able to get one for this semester, so I could try applying to Berkford. Or somewhere out there in California, anyway."

Freya saw the conflicting loyalties of her life struggling in her eyes. But then Alexis sat up resolutely. "I'm just spoiled by the sunshine in California and New Mexico!" she said. "And after this scary week on the road, I'm happier to arrive in London than I've ever been before." She smiled a determined smile.

**

"Message from my mother: they're coming to pick us up," Alexis said after they had arrived, through a staff kitchen parking lot, at Heathrow Airport. "She has invited you two to spend the night at our house, because it will be late."

Spend the night? Freya stopped short. *That's all?*

CHAPTER FIFTEEN

LONDON LIBRARY, THEIR FIRST DAY there. They had arrived at this sanctuary, a quiet place in a country where they could not be harmed. Ozzie hoped.

Here they could gather their wits and begin to execute their plan.—Which was a bit vague to him, since all they had been able to talk about, from the start of this journey, was how to get out of New Mexico, and how to keep going fast, and how to eat and sleep enough along the way to do that.

Now that part was over. So now, actually it was time to put a little more work into the plan. To think about the details a little.

But at first Ozzie couldn't think at all, and Freya seemed to be in a similar state, so they found a study desk they could share, far off in a remote area of the library where no one else seemed to want to be. They turned on the lamps and hung their daypacks on chairs that faced each other. They searched and pulled pulp-books off the shelves just because there were so many to choose from: her book was one on Icelandic seismology with large illustrations, and his was a book on the history of flight.

They sat and became engrossed in their books.

For a while.

"Freya," he whispered. "We have problems to solve. Just banging around in books makes me feel like we're going nowhere. We were going to try to figure out what Raker's up to, right? And talk about the volcano plan. What's the plan?" He felt for the ring. It was hot now, really hot.

She looked at him, a little uncertain. Actually, puzzled.

A new thought struck him: "You *have* no plan?"

"Well, *no*, but..."

"How can you have no plan? What are we supposed to do, then? We have no jobs, we're being hunted in America, we have no place to be for long, even here—" He knew he was probably being an irrational idiot right now.

"Shhhhhh, Ozzie," she said. "We'll figure it out."

**

Freya saw that he was not satisfied with her answer. In fact, he was coming unglued.

"What are we doing here in this library, anyway? Just

hiding out? Just getting out of Mrs. Wu's way?" He sounded desperately unhappy.

"Ozzie." She leaned across the study table and took his face in her hands. He pulled back a little, involuntarily, then was still.

She was so aware of how closely he was looking at her. She suddenly felt naked, sure he could see everything about her, every secret, right to the core. Maybe that was right, though. There must be a time for you to be this way to someone else: everything shows.

She wished she didn't have to speak. Maybe he understood? But his brow was still furrowed with worry. Finally she said: "Look at me. I'm not magic, Ozzie. But I know what you and I can do together."

It was all she could think of to say. So she stood and walked around the desk. She tugged his hand till he stood and then she kissed him. That ended up taking a long time.

**

Sometime in the middle of the kiss Ozzie realized something: Freya was like a witching wand.

Like those willow sticks that water-magic people used when they wanted to find places to dig wells. There were a few water-witchers out around Las Cruces, Malo said; they walked a piece of land with a willow wand in their hands, and when there was an underground stream or spring the wand would tremble and dip downward.

That was because it had affinity for the water beneath the ground. Or maybe because the water-witcher had the affinity.

He saw Freya: how she did this witching-wand trick. How she became the wand. She herself turned toward the right direction, as she went. She just went forward into things that she didn't understand, and kept on going until she knew which way to go. Now, he looked back and saw her doing it in Egypt. And on Mars.

I could never do that, he thought. *Makes me feel crazy. No structure. But Freya: crazy-haired Freya is expert at this kind of thing.*

While they were taking a breath, he whispered something like that to her, to explain what he meant when he finally concluded, "I can't do that, Freya."

She said: "And I could never do what you do, Ozzie. The way you plan many moves ahead, and bargain, and think of clever things to say at bad times, or things to be watchful about. And the way you see so clearly the fairness of things—"

"You do, too."

"...Yeah. Well, we're a lot the same, but a lot different too, huh? That's why we're such a good team." She took his hand again and he was surprised to find that he was a willing captive. His chest swelled to admit a little more happiness.

Instead of letting go, she kissed each finger of the hand, one by one.

At that he was lost, all his defenses gone; lost in Electric Freya-land again.

The bookshelves bent and rippled when he kissed her back; they became ancient, hung with vines and filled with runic texts, smelling of leather and roses and noisy with the blood of heroes.

Everything around them went magic. And none of that stopped until he heard something new intrude itself into their private world and clear its throat. It was wearing lavender perfume.

Ozzie snapped out of it first and froze. "There it is, right there," he said, pulling a book from the shelf behind Freya. He took her arm. They turned to the table and sat down primly to read, pretending the librarian wasn't nearby.

Until in fact she wasn't.

...She is too magic, Ozzie thought incoherently as he stared at a page of their new book—at one of 250 brilliantly colored illustrations of *Hats and Headwear Through the Ages.*

**

He didn't know it for sure right now, as they walked toward the Underground homeward from their first day at the London Library. But after many days had passed he would be certain: she had changed. Morning and evening, bent over a book, whispering with him and drawing diagrams, eating her lunch, walking among the pigeons homeward, as beautiful as if she had the wild desert moon of Egypt shining on her every minute.

Even the perfectly OK, everyday Freya was gone. And this radiant person was with him instead.

He wondered if he could be hallucinating. It was possible.

**

As they rattled toward the stop near Alexis' house, a message came in on his phone, from NoID, Diana:

Malo says Qualen had visitors at the Reed house yesterday: two men looking for your father.

The light in the Underground car, when he looked around them, was dreary. Ozzie noticed the cold and shivered.

**

Ozzie and Freya walked from the Underground station toward Alexis' house in the gray of dusk. "Good thing you had the idea for our parents to leave New Mexico and leave no information, huh? And Ozzie, I don't think we should worry that we're not so brilliant today. It's only our first day. Maybe we just needed time to get over traveling," she said.

Somewhere a clock chimed 5 times. Freya counted the chimes out, because it was something new for him: although it came up in stories sometimes, no clock in Las

Cruces ever seemed to do that. He stopped and stared at some shattered ancient surveillance cameras hanging limply from the light-poles at the intersection near Alexis' house. They looked trashy and old-time sinister.

Five o'clock: Which was ten in the morning in New Mexico, where they weren't. They had escaped; it dawned on him fully. *We did it. Made it to London.*

Back in New Mexico they were just missing persons. Qualen, who knew nothing about where he and Freya had gone, might be weeding the garden, after milking the cow and feeding the livestock. The September day would be warming there, in the clear air and sun, while they stood in the gray and damp here in London. And in sunny California? Norm might be getting up, barely.—Or, knowing Norm, he might not.

Last night they had been almost too tired to think, after the ridiculous flight and the wait at the airport for a family car to get free so someone could get them. But the wait turned out to be a lucky break, because they looked awful.

When Mr. Wu finally arrived, he and Alexis and Freya were as ready as they could be: only a couple of days away from showers, washed and combed and benefiting from the change to fairly unwrinkled clothes in the airport bathrooms.

Mr. Wu eyed them all skeptically, embraced Alexis, and didn't say much until they were driving homeward.

"Noticed that your semester has already started," he

said to Alexis. "But perhaps you feel that your advance credits put you ahead at the start?"

It was a polite question, Ozzie noticed, but not one that would put you at ease, either.

"It will be fine, Father: I have the scholarship authorization, and my credits for the Mars project *do* already put me well ahead. I have sent word of this small delay and there has been no objection...Do you think that Grace and Edward and Winston would like some rocks from Mars? I have also saved one, the best one, for you and mother... If you buy them, they cost a lot."

He nodded, politely.

When they reached the front steps of Alexis' house, a tall London row house sandwiched snugly between two others like it, Uncle Yong sat smoking. He sprang up and hugged Alexis eagerly. He pumped the hands of her friends, asking if they were robotics experts too. He was curious about New York City.

Spotted Tom had left behind his water-soluble cardboard flight box at the airport, and had actually obeyed Ozzie's order to be silent in the cloth shoulder bag all the way to the house. But while Freya described the Statue of Liberty to Uncle Yong, Tom asked (firmly) to be let out.

[London, a big city,] Ozzie warned. [Want to live in the back yard here?]

[Not worried. I can take care of myself.] Ozzie went down on one knee by a shrub, pretending to adjust his backpack, and Tom disappeared over a wall.

Later, when Alexis had showed Ozzie and Freya where the towels were in the Wu household, he said quietly, "I see you got us more than an overnight stay. Took cleverness, I'm sure. Thanks." He realized there were things Alexis couldn't tell her mother: why they couldn't send word ahead, that they needed long-term lodging, and that they were being chased so they needed safety. And he hoped all of those would never come up.

"Yes," she whispered. "Now you're safe from Raker. You'll be able to get a lot of research done each day, in the London Library."

"We will?" Ozzie hadn't gotten that far yet.

"I told my parents so."

He got it. "Then we will."

CHAPTER SIXTEEN

"SHE WON'T LET US WORK?" Freya said. "Why not?"

"My mother thinks it's wrong for guests to work."

"But what if we're here for a while? More than a week or two?"

Alexis looked wary. "Don't know…"

It sounded to Ozzie as if their stay here was misfiring, but they hadn't worked out any other plan. Yet. "Hey, Alexis, what if we do your chores while you get caught up at school, to give you a fast start? Would your parents allow that?"

She sighed with relief. "Good idea, Ozzie. Yeah, they'll buy that, at least for a week or two."

**

After Mr. Wu had left for work, whatever it was, and Mrs. Wu had gone to take the excited children off to school, Ozzie and Freya cleaned up and moved fast to get the first items on Alexis' chores list done, sweeping the kitchen and washing the dishes, before they left for the Underground.

They decided a few things on the train, while it slid onward like scissors cutting, metal against metal, and they watched the Londoners entering and exiting. Ozzie announced, "I'm not real big on research. I've mostly researched things that you look up once and you're done, like planetary atmospheres and metals prices, fuel usage, engine capacities...What do you think we need to research next?"

"Space companies for you to apprentice with," Freya said.

"OK, that'll be fast," Ozzie said with a shrug. "It will be hard to find any that are big enough. But about the volcanoes, while we're here we'd better find out what's really going on: what's causing them, and what Raker is hiding, right? And what to do to fix it."

"Right. Well... Remember in Giza, Raker wanted us to know that fracking isn't something we need to know about, or electronic waves? Waves of energy, waves of sound; he was *sure* we shouldn't care about all those things. So maybe we need to know more about those, right?"

His eyes flashed with rebellious joy. He looked like he'd do anything for something real to fight, right now. Ozzie slapped hands with her. "Let's do it."

They were at the front door of the library when it

opened. They found the same table in the back and went to work.

**

Maybe Raker is helping us again, Freya thought. *He just can't keep himself from doing it, maybe.*

So far he was definitely telling them they were on the right track, by telling them that waves, including sound waves, were just what he didn't want them to look at.

Also, the things he had told them they should not be concerned about seemed to be related, and there were many variations of them, all involving waves of energy or sound or both—waves going through air or ground—and all of them were possible sources of disruption for the crust of the earth.

If some technology like the formula in Ozzie's ring was being used to cause destructive sound vibrations in numerous locations, the vibrations could be disguised as manufacturing, or weather control, or fracking—anything.

Now they had more places to start looking, anyway. *Who said that? About your enemies helping you?*

**

The trouble is, Ozzie thought, Freya's already gotten this far before, without the London Library.

Piles of books and a heap of data slivers surrounded them on the table. They sat side by side at the only cleared place, with one recorder playing data slivers on the screen and the other taking data requests.

She was drawing a diagram of the way layers of porous rock on the earth, which contain water, could carry vibrations easily because of the water in them. Like the pyramids, whose "feet were in water," as their Egyptian guide had said.

"So," Ozzie said, "someone may have discovered that sending heavy vibrations, or continuous destructive vibrations—like ones caused by the sound formula Diana found in the ring—into these sorts of layers would transmit and magnify the vibrations."

"It seems true. And they could gradually shake the crust so much it cracks."

Right. So hot magma from below could pour through. He imagined again the scenes she had described: the roaring flames she had fought in Iceland, and the terrifying lava flows torching pastures and killing herds.

He leaned back and raked his hands through his hair. "Well, if vibrations are causing the volcanoes—and you guessed that was true long ago, Freya, didn't you?—they could take the form of any of these things we've been looking at. But which one of them, and why—"

"—Yeah, why? It looks like Raker is hiding things about the volcanoes, but even if we make guesses and say 'OK, Raker and GG must be doing something insane to cause a volcano in Oklahoma' we can't prove it. And it still doesn't tell us: why would they?"

Ozzie shut his book in disgust. "Doesn't tell us how to stop it, either."

**

Ozzie was texting Diana again when Mr. Wu called them for dinner. Ozzie finished up as he arrived at the table. "To ask her how things are going," he explained to Alexis' father.

"Very proper that he inquires after his mother so often," Mrs. Wu said. She added to Alexis: "Maybe you should learn from your friend?"

Alexis sighed and tried to stifle a lopsided smile. She busied herself getting seated next to her younger sister.

**

In the morning Freya did an online check: she looked up Ilse Arnsdottir in Iceland. A local posting on the worldweb said:

Local Talent Returns to Town: Ilse Arnsdottir will be singing next weekend at the Bryggjan with new husband and singing partner Doug Reed.

Ozzie nodded when she showed him. They were really there, and safe. Or safe enough for now, anyway.

**

A text came in from Norm, with his wise-guy grin

above it. Freya missed him, suddenly.

Ozzie summarized the message from his phone: "Norm says he was followed around Berkford for a week before he caught on. He began wearing fake wigs and other disguises to class. Look at him." Ozzie grinned at Norm's holos of himself in costume and tossed them into the air between them for Freya to see. "Says no one seems to notice."

At the same time as she did, he thought the next thought: "We'd better start wearing our disguises again too, Freya."

**

Freya looked up from the holo she was prodding. It hung in the air at nose-level, turning and unfolding like a nasty flower: a cross-section of a volcano forming and erupting.

Ozzie sat opposite her, at their latest desk among the shelves of odd-sized books in this remote part of the library, with the thin arc of the earphones hung around his neck. Books and his player and holo-slivers were piled around him.

Right now, he had told her, he was learning about how to damp sound waves down—how to slow them down and make them smaller—as a possible cure for the destructive waves that were disrupting Earth's crust. But after he punched the button to start again, she recalled: she had already learned a year and a half ago that no technology for damping down waves had ever been successfully used on a large scale, like planet-wide. Nothing even close.

He hunched forward, his finger searching the page eagerly. For a person who didn't like research much, he put a lot of intensity into it.

While he was here to help, while they had the time to look, she had to find the next place for them to look for some real answers. For something they could do.

He sighed and turned a page, scowling a little.

A minute ago his eyes had been clear blue-gray; but right now thick lashes screened them from her. That was a good thing; she wouldn't want him to catch her watching his eyes.

She thought of two years ago, when she searched alone for answers, and had no London Library either. She was lucky to have him working with her. No time to waste. She put another data request into her player.

CHAPTER SEVENTEEN

AFTER DINNER A FEW NIGHTS LATER, and after Ozzie and Freya had finished tutoring-time with Grace and Edward and Winston, Alexis left her three siblings doing the dishes and joined Ozzie and Freya outside in the courtyard to hear Diana's latest:

> You know PII has been trying every way we can think of to use the Singer's voice for the most powerful effect on the vibrations.
>
> So far we have not been able to amplify her voice electronically and make a huge difference in the vibration. If we amplify to make the sound twice as loud, the power of the music gets greater by about 2 times. If we amplify to make it 3 times as loud, the power seems to get greater by about 2 and a half times... there's only so much we can amplify it before it breaks eardrums and makes no more change at all.

But we're finding there's a big difference with the addition of more people. That makes more difference than adding sound volume electronically. Double the number of people singing and the effect goes up by about five times!

That's good news and bad news. Bad news: that it can't just be done electronically. Good news: it can be done anywhere there are voices.

Alexis said, "Just like my dream, right? At our Mars greenhouse," she added smugly.

Freya couldn't quite remember, but she nodded, thinking.

**

"You know what, Freya?" Alexis said, next morning at breakfast: "I had a weird cat dream. My sister's cat, who won't even *look* at me, usually, came and..."

Her mother turned toward her quizzically.

Alexis caught herself. "You know, cat pestering me at night, Mother! And I woke up with the old thing sitting practically on my head!"

When Mrs. Wu left for a moment to call the younger kids, she whispered, "I remembered Egypt and I tried to recall if there was a dream. Right: I did dream, just before I woke! There was a choir of animals all in a greenhouse singing, with the ground shaking, everything roaring. They sang and the noise all around us got quieter; slowly the ground finished shaking. I swam through dark water and saw greenhouse lights, far away, turn into a net made of garlands of greens, all hung with lights, holding a ball of blown glass: light blue, in a live net of green...

"Weird, right?"

"Yeah, pretty weird," Freya murmured back, imagining how that must have looked. But there was no need to be quiet now; the others were clattering into the kitchen by then.

**

"Well, I have news," Alexis announced that night at dinner. "I've joined the Oxbridge Chorus."

Both her parents looked startled, so Freya guessed that this was something very new.

"Well, you *know* I'm not a fabulous singer, but the Chorus is for anyone who wants to sing for fun, you see, and I thought it would be fun..."

Ozzie's fork was poised near his mouth, and his mouth was slightly open, half-smiling.

He's thinking, Now what, Alexis?

"...And," she added, "I have a piece of music I want the Chorus to try out. Very unusual piece," she explained to her parents; "some... strange sounds, from Mars, that we used in our greenhouse. I've called it 'Mars magnetic music,' told them it was difficult to master—and that hooked them. We had such fun today struggling with the notes! We are rehearsing it now to sing at Michaelmas, at the end of the month."

**

On her phone Freya read down through Alexis and Norm's conversation, copied to her and Ozzie: "Crazy busy here, Norm! We are so busy practicing!" Alexis had sent Norm the latest recording of the Oxbridge Chorus, 100 strong, doing "The Growing Song."

"Sounds corny," he said back, live-holo, with gestures and facial expressions.

Freya couldn't help chuckling. What a clown.

"No soul!" He ranted. "Play them some blues or something so they get a better idea of how it should sound."

**

A week or so later, the Oxbridge Chorus performed their idea of Ahanith's Growing Song, with a certain mournful Memphis twist to it. Ozzie and Freya came up with excuses, for Mr. and Mrs. Wu's benefit, for not going, but Alexis promised them a full recording of the performance.

"The audience was fascinated," Alexis said, when she sent the recording of the song and its performance time off to Diana, with copies to Norm and Ozzie and Freya.

Diana checked PII monitoring records against the

performance time and told them all it made a difference in the worldwide tremor recordings. It was a small improvement, but actually bigger than electronic boosting *or* Malo's gypsy group of 85 singers had been able to make. It was right there in the records: a change that was measurable and timed exactly with the performance clock.

**

"Ozzie, look," Freya said sadly. "Raker:"

Today United States military personnel began to move into the Cherokee Nation in Oklahoma, to enforce yesterday's U.S. legislative decision that the Nation is incapable of managing its citizens through the difficulties caused by the Oklahoma volcano. "An emergency managing government sent by the World Council will be installed immediately, with the cooperation of the United States government," said Representative B. Arnold Raker.

CHAPTER EIGHTEEN

FREYA WAS OFF AT THE NEARBY GROCER'S stand; her turn to get them some things for lunch. She would holo when it was time to meet her out in the park.

A newsfeed came in about Doug and Ilse, so he stopped the player and opened it on his phone: a worldweb review on the latest place where they were singing in Reykjavik.

> **At the Kaldi, Ilse and her partner now serve up punchy, saucy duets that are bringing worried Icelanders out of hiding for a night of music and good food...**

He smiled. Then sighed, and poked to restart the player. He had an audio sliver playing in his earphones while he looked over the pages in a book at the same time. Ozzie was sick of sitting these days, so he compensated by sprinting through stacks of library knowledge as if mental speed itself was his sport.

As a result he chewed through information hard and fast. He knew he'd better: while they had the chance, safe here in London. While the Singer tried each day to sing the planet to the end of its quaking—and while PII learned from that. They needed to be ready when more data from PII would open new research doors for them. *Wish Freya and I could open some new doors, too.* He felt doomed, and speed was the only cure.

He could hear some kind of rumbling all day every day now, below the traffic noises in the London streets, at the Wu dinner table, even here in the quiet of the library—and it was no use wondering what was causing it any more.

He was urgently motivated, without a doubt; just like the days when he was learning about robotics to catch up to Norm. Now he was learning about energy waves and sound, trying to catch up to the volcanoes. He stopped to stare when he realized that this was what Freya had been doing ever since the day he met her: racing against some weird volcanic clock.

Today he raced to find out about differences between live and recorded sound. And he was just beginning to get somewhere, when a very fat chocolate-brown cat suddenly arrived in front of Ozzie's face, all four paws planted firmly on the book.

"Huh? How did you get here?"

It sat down on the pages and began to lick one paw

carefully. Ozzie looked around for a cat-carrier, for a way a cat could enter, or for another person who might own this cat. No one, nothing.

The cat did very detailed work, but eventually it changed paws and began licking the other one. This cat's way of starting a conversation was unusual. Finally Ozzie said: [Hello.]

The cat paused, gazed at him for a few seconds, then began licking again. As she did, the bookshelf shadows on the floor seemed to warp and bend, and out of them oozed several more chocolate-brown cats. Somehow their color was perfect camouflage for this place.

The new chocolate cats seemed to ooze uphill effortlessly to stand in front of him on the desk.

Again he looked for the opening through which they had come. (*Cats? in the London Library?*) His jaw dropped: now a half-dozen more were oozing out from behind the tall shelves and becoming a thick chocolate river up onto his desk, crowding tightly together. Several sat on his book, and the others packed around them: in a variety of chocolate colors, like a box of candies.

[Hello,] he said again. No answer. [Good day,] he tried; maybe British cats would do better with that greeting.

Some of the cats shifted nervously. No one answered.

[Something you want to tell me?]

They all squirmed until the first cat hissed at them with quiet menace. The group went still again.

Ozzie was baffled. They seemed to have something to say, but they didn't know how, maybe? Not living with gypsies, maybe they weren't so practiced at talking to people? Maybe the Brits thought the cat-voices were their own imaginations and refused to talk to them.

The first cat hissed noisily at him. His former house-cat, Georgie, now a resident of Mars Colony, had done that sometimes: not a fighting sound, just irritation.

Ozzie settled back in his chair and gazed into the eyes of the cat for some time. After a while he thought, as clearly as he could: [I am listening. Tell me.]

He heard a faint voice shout, as if it came from half a mile away, [MARS!]

He heard other voices now, a group of tiny, faraway shouts echoing after the first: [MARS!] Yet the shouts seemed to be coming from the cats right in front of him.

Ozzie's phone sounded. He slid it out. [Excuse me,] he said to the first cat. "Freya?" he whispered. "There are a bunch of cats here talking with me about Mars... No, I don't know. We just got that far: Mars."

[Um, can you meet us in the park?] he asked the first cat. [Third bench?]

He reached out to stroke the cat, experimentally. His hand went right through it.

He left the cats standing there on his things and went outside to meet Freya anyway.

**

There were twelve cats already at the park bench when he arrived, although most of them had curled up in the shadow beneath so they were hardly visible. Two cats sat directly on the bench beside Freya and their lunches. Ozzie reached out to stroke the nearest one, experimentally again. Its fur was cool and glossy, as real as any cat fur ever was.

Freya talked with the loudest one:

[Know Georgie?]

[Who's that?] It was a very faint shout.

[Know Banzai?]

[Not even a little!]

Ozzie looked at Freya and shrugged. She said, [So you want to go to Mars. Why are you asking *us* to do this?]

There was a pause, a faintly embarrassed one. [You seemed to be paying attention.]

Ozzie gave Freya a look. "How are we going to get them there?"

"I think the Icelandic Consulate will help us," she told him. It sounded like it could work.

Ozzie hated to stop researching right now when he was running into some good stuff. But the cats should go, of course.

The woman at the Consulate desk was happy to chat with Freya in Icelandic. She was apologetic, though. Freya translated. "We can't ship to the Mars Colony because Iceland has no base there, or vehicles for space travel yet.—Wish we did," she confided. "This shipment must be arranged and accepted by a government that has the ability. Like Britain."

They were wearing the best disguises they had, and none of their ridiculous road stuff, the blonde wig and the old hats. These days, Freya wore heavy makeup and dark curls, a real wig from a thrift store that fit well. And Ozzie used a ski cap and eyeglasses with trendy, distracting rims and lenses that were just for style so they didn't make it hard for him to see. Their clothes were nondescript. No one would pick them out in a London crowd.

The cats had willingly crowded, six to a box, into the sturdy boxes Ozzie and Freya had scavenged from the alley behind a bookstore. He guessed they'd better get them shipped now, while they had them all boxed up. So they sat on a bench and called Alexis.

**

With Alexis they walked the last few blocks toward the British Foreign Office. The box was getting a little heavy, but Freya wasn't going to mention it.

When Alexis had arrived without a disguise, Ozzie was alarmed. But Freya decided, "She is a British citizen, after all. I guess she should be safe here." Now she noticed another of those old cameras, like the ones in New York City, attached to the cable of a signal light. "Do those work?" she asked Alexis.

"Not for years," Alexis said. "When I was about ten

there was a big beautification campaign to take all that junk down in London. My school was given a day to help with it. That one up there was probably too hard to take down."

Freya didn't like it there anyway. Just to cheer herself up, she said, "Did you hear that the Cherokee Nation people have posted guards on the roads? So no one can enter their land to take over?"

Alexis led them into the British Foreign Office Mars Commission. She stated their business in her most professional voice to the receptionist, a thin young fella with a reedy voice.

"Is this some way to get rid of cats? Are you here from the Royal SPCA?"

"No," Alexis said. "These were volunteers." She suppressed a giggle.

"What? Cats? Volunteers for what?"

"She means, chosen for the purpose," Freya said quickly. The receptionist stared at her. Maybe her accent showed up a little. They could be identified and located if they drew attention here. Unlikely, maybe—but right now she had a bad feeling: danger.

"Yes, check with the British Mars Colony," Alexis said. "They are for Clara Martin, Cook, from Alexis Wu, shipper."

As Alexis signed the papers, shipping to be paid by the British Mars Colony, an automatic photo device flashed, taking a documentation photo of all three of them plus two large boxes of cats.

**

"Don't look back," Freya said. "But someone is following us." She'd seen the reflection glance off a shop window.

Immediately Alexis, who was in the lead, led them left into a store entryway, pointing at a new style in the window display. They pretended to look at shoes as they watched him pass: a man in business slacks and a jacket. He went by them facing front, but Freya saw his eyes slide toward them.

CHAPTER NINETEEN

LEAVES SCUDDED DOWN THE LONDON STREETS as they walked to the Underground these days. When the Wu kids caught Ozzie and Freya in costume by accident—they had just put them on at the back courtyard gate when Winston ran through it—Freya was able to cover for them by saying "it's nearly Halloween," and making up a story about a costume contest that they might be holding at the library.

On the Underground, Ozzie's phone sounded: news from Diana. He showed Freya.

> **By the way, I've given a master of the Growing Song to Doug and Ilse. I asked them to sing the music there. They've added it to the end of their performances in Reykjavik, getting the audience to join in on Ahanith's song. Here's a soundbite from your father and Ilse:**

Ozzie was happy to see their faces. The holo showed

the two of them chatting to Diana at a mike.

"We picked the notes that would be easy for U.S. and European people to sing," Dad grinned.

"And gave 'em Icelandic words," Ilse added, "to support the Icelanders."

He hit a guitar note and together they crooned a part of the Growing Song that Ozzie recognized, in their own performance style: a country-music duet. They took their harmony way up to the high notes, sent the sound looping through the air, and brought it back to land in the kind of sweet finale he had heard them do before, evenings at the house.

Freya smiled. It was good. Catchy. A crowd-pleaser, for sure.

Ozzie wondered if it would do anything much to the volcanoes, though.

**

"Damn." Ozzie slapped the library tabletop in frustration. "I wish we had access to live singers to experiment with, like Alexis."

While PII's private investigator was trying, way too slowly, to find out more about the Oklahoma companies, Ozzie and Freya had decided to look into the reason behind Diana's discovery: that if you doubled the number of live singers instead of boosting the volume, you could reduce the shaking of the Earth's crust five times as much.

"Yeah, me too... Well, maybe if you and I can get an idea about why it is that live voices work better, PII and Alexis can try the idea out for us."

So they studied the wave patterns of live voices and recorded ones. The patterns seemed the same. What made the live voices different?

"It can't be emotion, Ozzie," she countered. "Emotion is contained in recordings as much as in live music, and people are moved by both live and recorded music, aren't they?"

"True," he said. "Besides, emotion might be what *people* care about. What does the geology of a planet care about?" They tossed holos up into the air between them to study, with the sound low, so they could follow one idea after another to a dead end.

"Maybe it *would* help if we could go to a recording studio and try things out," Freya said finally.

They were sitting on a park bench near the library, pouring out their lunch crumbs for the pigeons, when Ozzie called the PII sound engineer to ask what he knew.

He didn't know enough.

He said, "Current instruments and tests deal with certain qualities of sound, but other qualities must exist that haven't been identified or measured yet." He was researching as fast as he could, he said; he also looked forward to hearing more about what they learned.

Ozzie shook his head. He certainly hoped *someone* would learn something soon. Something useful. And why was the PII private investigator taking so long? Most of all they needed his information, and maybe just a little more time…

**

"You're becoming a clever researcher," Freya said the next day.

He didn't think he was doing too badly either. Between other things Ozzie had been throwing in some career research, too, mostly to keep Freya from worrying about it. Today he said,

"OK, Freya, I've found five New Mexico spaceflight companies, and two Oklahoma ones, for my list. That seems to be it: those are the only ones big enough to have trading lines and use space-traders. …So now that I have a list, I can go there and apply anytime, right?" Ozzie couldn't hold back the edginess in his voice. "While we are hiding out from Raker?"

"Yeah, I know," she answered. "Useless to go back to the U.S. with Raker causing trouble. Well, I guess we just have to expose Raker as an idiot before we go back there." She managed to make him laugh by doing Raker imitations: Raker at the Giza jail, Raker on the news, Raker's opinion about the blahblah volcano…

When they stopped laughing, Freya said: "It's a good thing that Raker doesn't own *all* the spaceflight companies. In fact, I'll bet some of those companies would *love* to take GG out."

Ozzie hadn't thought of that. GG had been his standard of excellence for so long. It was hard to think of the company as the opposition. "True. Raker can't stop GG's competition from hiring me."

"That's right." Freya's eyes flashed loyally at him.

"I guess, while we're making an idiot of Raker," he grinned, "I can at least introduce myself to these other guys and try to get things rolling..."

He wrote up job applications and a letter to go with them. His letter said only that he had outstanding references.

Freya offered to help with the wording because his version seemed too matter-of-fact to her. She wrote: "My student record and commendation from the British government are attached. I was first in line for the Grand Galactic Captain's Apprentice award, until the son of a U.S. Representative, Seth Raker, entered the competition. When he did he was given 'most qualified' status, in spite of his record of cheating and physical abuse. If you'd like my help to make your organization more competitive with Grand Galactic, I'll be glad to assist."

He grinned and signed it electronically with a flourish, attached the letter and documents together, and hit the button to send it to the seven companies from a protected site. Even if there was no other payback for this little project, there was the one right in front of him: Freya's eyes sparkling at him with ornery delight.

**

"Why is the private investigator so slow?" Ozzie asked Diana in a phone message, while he and Freya sat eating a wind-blown lunch that day. "We need to know who's doing what, so we can stop it!"

"I wish the investigator were faster too," she sent back. "To be effective, he must stay unknown—so he has to be cautious. He may have taken a local job as a cover."

So Ozzie volunteered: for the next few days, in addition to sound research he and Freya would help by researching Raker's history and connections, because London Library seemed to have more extensive resources than the private investigator did. What they learned they would send to PII just to help out.

"Ohhhh. Ozzie, look at your newsfeed."

He did. His lunch turned to stone in his stomach. There was the photo of Alexis, Freya and him at the British Foreign Service Mars Office, standing with their boxes of cats.

He groaned. Raker's disgusting news story told all about it, and they could read between the lines: Alexis was being followed daily without knowing it. Just like Norm was. Stupid of them not to guess that! Of course Raker could afford his own British private investigator.

So when Alexis showed up in a public place with two strangers recently, someone had gotten an OK to go through the day's security system photos at the British

Foreign Office. After that, comparing the security-system photo to other photos of Freya and Ozzie wasn't hard.

"They can't *prove* it's us, with those disguises!" Freya said. True. But they had been located again, he thought. Raker could get someone to start hunting around London for them.

If not enough progress and too much library time were the worst things about their lives right now, here was something that topped both of them easily: Raker now threatened extradition, demanding that the British government return the two of them to the U.S. for trial.

They could be spotted anywhere, after those photos.

Like right here, for instance. Ozzie quickly pulled off the telltale hat and glasses. They left the windy bench and slipped to their desk in the library by zigzagging through the shelving most of the way. They hauled their research materials to a new table, equally far from the entrance. And they sat warming their hands unhappily.

"The British government wouldn't listen to him about the extradition, would they?" he said, to encourage himself as well as her.

She shrugged. *Don't know.* She looked as anxious as he felt.

To be made into a public enemy in the newspapers made him fume. No one should believe such a stupid story, but someone might. And you could count on Raker to make it world news. Who would pay any attention to his job applications if they'd heard this extradition stuff?

They needed something to cheer them up, but even a visit to a coffee shop wasn't going to happen now. Freya found them some more oversized picture books to look at, and sat turning the pages of hers.

When Freya was finally lost in her book, *Northern European Sailing Ships*, Ozzie hid behind his own pages and started researching escape routes, quietly, on his phone.

**

Within 24 hours Raker's story was in all the newsfeeds and getting lots of exposure. Some of the feeds showed student photos of both of them. Raker had influenced someone in Parliament to support his demand for extradition of Ozzie and Freya to the U.S.

"—*For being unpaid research assistants on the successful British Mars project? Is this why Britain should spend its time and money extraditing them?*" Freya wrote in a dozen letters to the editors of the London news services, under the name Gerald Freeman.

"If I've ever hated anyone, I hate that man," Freya muttered bitterly on their way home from the library. She had added a hat and dark glasses to her disguise. Ozzie had put heavy gel into his hair to make it dark, and wore it falling into his face.

He nodded. His fingers went to the ring on the chain at his throat. Hot again.

**

Ozzie overheard her father, who said, "We have treated your friends with respect, as the people who helped you on Mars. But now..."

"Now we will dishonor ourselves if we are unfriendly to them," Alexis said. "I was a guest in his father's house for a week, then a week plus more days—"

"Yes, and they have been here—"

"They saved my project. They saved my life." Rapid-fire, she told the story of Ozzie and the trucker who almost ran away with them in New Mexico.

"You were hitchhiking!" her mother was aghast.

"That's beside the point, Mother—"

"It is not. You should behave like a student, not like a vagabond."

Her father added, "And friends with disguises...are these the friends you want?"

Later, Alexis sorrowfully apologized for her family to Ozzie and Freya. "Please try to understand. The news people make it sound so terrible. They're just afraid, that's all."

Alexis sat on her bed and watched them as they loaded up their backpacks. Ozzie was sorry for her. "It's OK. No one understands parent-drama better than we do," Ozzie said. Although hers was definitely worse. "Right, Freya?"

"Yeah," she nodded.

He guessed that Freya didn't sound British enough yet to go unnoticed. "Yeah" was still the biggest place where her English was laced with Icelandic, but it probably wasn't the only place.

Well, he couldn't see any other way... And just when he thought they might be getting close to something, there at the London Library, too. "If they're worried, it's time to go."

Alexis' face was very pale. She said mournfully, "You know I am *longing* to go with you. But I can't..."

They knew.

"Please tell the kids that we'll be moving to a new library to research," Ozzie said to Mrs. Wu at the door before they left. "In Portugal."

"They may tell their schoolmates—" she warned.

Good, he thought bitterly. *They can tell all the neighbors, too. Because we won't be in Portugal when Raker looks.*

"Will you keep watching out for Tom?" Freya asked Alexis on the front step.

"Of course. He seems to like to travel a lot. I'll tell you when I see him next. I'll walk you to the bus stop."

The bus disappointed them all by arriving right away. They didn't even have a chance to take their heavy packs off for a minute. "Stay warm," Alexis said. Her own cheeks had gone rosy from the chill. She stuffed a pair of rolled-up sweaters into Freya's arms.

They climbed aboard. They waved till she was out of sight. A few blocks later, Ozzie watched Freya unroll and look at the sweaters.

"Her best ones," she said a little sadly. There was some money rolled up inside one.

CHAPTER TWENTY

NORM TOOK THE LATEST, MOST DIFFICULT route back to his dorm room from class. He walked to the student building and into the coffee shop, then out a back janitorial door and through a private garden adjacent to some administrative offices. He slid through the space between a hedge and a wall to return to the main walkway just outside his dormitory door. A quick look around showed no one in sight who seemed interested in him. He did the front steps fast, ran up two flights, and locked himself in his room.

Now he could look at the text from Alexis.

> **Ozzie and Freya may be forced to leave England. Raker has succeeded in making a royal fuss here. Asking for extradition. My parents were horrified to find that the two assistants who are here with me are illegal!**

He groaned. He hated being here away from the

action, where he couldn't do a thing.

8 p.m. in London.

He called.

**

There she was, live-holo, sitting cross-legged on his desk:

"Raker is really campaigning against us four. But he can only get at Ozzie and Freya right now. It says, 'Conspirators to be extradited.'" Norm rolled his eyes and punched the word "extradited" into his phone.

"So good to see you using the dictionary, Norm." Alexis smiled sweetly, lopsidedly, but not too happily.

"Thrown out of Britain?!" he was astonished.

"Turned over to the U.S. for trial.—Well, that's according to Raker, Norm." She read on: "'British authorities declined to comment.' So maybe they'll just tell him to buzz off. But you see, they've left London. Ozzie and Freya. They had to, because my parents were too frightened to keep them here."

He snorted, but he felt bad for Ozzie and Freya. *Where are they now? What am I supposed to do about it?* "Alexis, too hard to think right now. I have a physics exam this afternoon..."

That wasn't even half of it all. He was still being followed by someone. And someone seemed to be checking up on him with the Berkford University administrative office, because for the last couple of weeks they had been calling him in every few days to ask prying questions, as if they were suspicious. Creepy questions.

Alexis made a sympathetic noise about the exam.

About the exam: actually he had an A, easy. But Physics, which so far seemed to be all about all the things that can't happen—you can't make matter, you are a slave to time, all that stuff—how did they know. *Who cares, anyway?*

"Let me get back to you later," he said.

"Promise?"

"Yeah." He clicked her off before he said enough to really start trouble. *Better not to worry her,* he hedged. He was confused. He switched on a video game and began blowing things up.

**

He waited till after the Physics exam, but not so he could study for it. That just gave him some stall-time to think about some things. He played video games for a couple of hours, very refreshing, then he returned by reversing his new route back into the student building, and slipped out in a crowd of basketball players to get to the Physics building.

The math was easy. By the time the exam was over, he felt even more refreshed and ready to figure out the complicated things, like what was right to do next.

He slunk back into his empty dorm room with a

takeout supper from the dining cluster and settled in to think. His roommate had dropped out of school two weeks ago, so it was too quiet. One good thing the guy had done: he'd taught Norm about karma. Now he put on some vintage heavy metal sounds to help him think. He turned off his phone.

He thought about the last year, Egypt and Mars, and his travel time to California from New Mexico. He had learned more in those travels than from doing most things in his life. What did he know last year? How to do pizza and robotics. Really that was all.

Because he had brought this topic up, he now reviewed things he had learned:

If someone helps you, you need to help back. Sometimes you'd better help in advance because you are gonna need it sometime.

People are not mostly jerks. Surprise, they are mostly good eggs. Trouble is, the real jerks always pretend to be good eggs so it's hard to tell sometimes.

Don't be careless about karma. You need to pay attention to that. Don't just blow it off when someone needs you. Take care of people. And things.

Women are complicated but they love you. You need that.

Stand up for your friends and your family. They're the ones who deserve it.

He ate for a while, considering. While he thought he napped a little. By the time he woke, he knew where he was going with this.

Ozzie and Freya were in trouble.

His student advisor had accepted his whole proposal to get school credits for the Mars work. So first semester credits were covered already. He was already piling up 2nd semester credits now.

He had gone home twice so his mom could feed him and coo over him. Such great food! The aunts and uncles had plenty of Norm-time under their belts. He wasn't expected home till holiday break.

What better way to shake off whoever was following him than to be missing entirely?

Not to mention: what about Alexis. What better way to get to see her soon?

Ergo, the solution is obvious.

He decided.

Before he could change his mind, he texted her: "Hold on. I'm coming."

**

Norm ripped off a proposal for "independent study in robotics, details to follow." He printed it and got it under his advisor's door at midnight when no one would be around to follow him, he hoped.

He loaned his dorm room to a school friend who hated his real roommate—woke the guy and made him come talk about it, making sure the grateful fella understood this was a genuine favor with just three paybacks to the deal: "Don't let anyone touch my equipment, tell anyone who asks you that you don't know *where* I am or *when* I'll be back, and holo me with your photo if any non-junk mail comes."

By two a.m., his Egypt backpack was stuffed full again and stocked with 35 protein bars, flashlight, knife, and water. He had downloaded some new games to his phone. At the last minute he stuffed his disguise in too. And into a cargo pocket went his entire stash of birthday cash, so his new roommate wouldn't be tempted to borrow it.

He cleaned his supply of olives, squeezepaks of juice, and tekryl-canned food out of the shared food storage in the dorm kitchen. It all went into a satchel. He snagged one of the kitchen can openers, solemnly promising the silent kitchen that he would make restitution later. Rides would probably be easier to get if he came with food instead of empty-handed.

Hitchhiking to London: ridiculous. Not safe, everyone knew that. *But OK, here goes.*

His back ached as he pulled the pack on. It still wasn't recovered from his dive off the farm truck a couple of months ago. But his complaining muscles eased up again as he moved; he walked for a mile under the stars to the edge of the campus, then faced the oncoming traffic and put out his hand.

CHAPTER TWENTY-ONE

NORM SAW ON HIS PHONE that Ozzie had received a text from Alexis, which she called "Norm is coming!!"

He read her version of his life:

Norm has finished writing up his Mars robotic project for first semester and meanwhile taking extra-credit robotics classes and Icelandic as a language-study requirement. He's been doing beautifully at Berkford...But he's coming to help!

This was Norm's third ride. So far the truckers had been great.

Only the gas-trucks would stop to pick you up. He had learned that. And they were usually old and decrepit, like Ozzie's truck or The Heap back home in Pasadena. He looked the drivers over pretty carefully, and he could tell that they were doing the same with him.

The first ride, from Berkford south and then across Southern California to an experimental artist colony in Quartzite, Arizona: the driver was happy to eat his food and take him along. Norm told the guy funny stories to keep him awake, then slept a little, then let the guy teach him more about truck-driving on the flat parts. "OK, you drive for a while," the guy said. "Stop if the road goes uphill, and wake me." He looked desperately tired to Norm.

The gears were complicated, but he liked that. He ground them less as he went along.

The second ride was a friend of the first guy, someone the trucker holoed when they stopped, named Purdy. Purdy liked Norm's food just fine but mostly he wanted Norm to drive from Quartzite eastward toward Phoenix while he slept. Norm drove carefully, amazed that the guy would trust him with his truck and cargo. *Maybe the first guy passed on the word about what a good driver I am.*

Purdy did the hardest part—drove them into the afternoon traffic around Phoenix. He had to go north from there to Flagstaff, and he had no friends to recommend, so Norm got out somewhere south of downtown Phoenix and put out his hand again.

The guy who stopped next looked like he needed a shave and a bath and lots of sleep. But his face seemed friendly. He asked a lot of questions. He was headed for northern New Mexico, he said, but he added, "If you know how to drive truck I could stay on 10 long enough to get you a good ways along."

"OK, you're on," Norm agreed, climbing in. "Want a protein bar?"

"Name's Sam. You're—? Watch for my cat there."

**

Cold air woke Norm. But it was sweet, full of fresh, familiar smells that he didn't quite recognize. The driver must be getting out for a break. There was a warm blanket on Norm's lap. *Did I bring one?* Puzzling about this, Norm fell asleep again with his cheek against the cold side window.

The water rose and fell, and the huge fishnet spread out as far as his eyes could see, floating outward to cover the water.

At each junction of the net, a light glowed; as the sun slid below the water and darkness came, the net became a web of lights floating on the water, lighting the surface for miles.

There was music coming from the water, or the net, like singing.

The net rose and fell with the waves, and lighted fish jumped up through the gaps in the net, sang and fell back through the gaps into the water.

He shook the net, like a blanket...

A cat yowled.

"Hey, watch out for the cat!"

Norm sat up and shook himself awake. "Sorry. Dreaming..." he said. "Sorry, cat. Hey, Sam. Want me to drive now?"

**

Norm woke, but he didn't know where he was. He was in motion. On a space liner? No, a truck: his head, leaning against the frame, was vibrating to the rough rumble of the road, which must be very patchy right here. The time display on the dash said 3:14 a.m. He had dreamed, maybe, because he remembered a voice...But the voice began talking again:

"Current statistics show that Americans sleep an average of 6 hours per night, and they only dream about one hour of that time. Such a short period of nightly dreaming is abnormal in comparison to worldwide statistics, but it has become normal for the U.S. and Europe. Our guest tonight, noted sleep researcher Dr. Giuseppe Shing, will speculate on why that is, and also on an exception that has surfaced locally: the average New Mexico resident today sleeps eight hours per night and dreams for an amazing four hours of that time!"

The guy rambled on in a talk-show-host voice:

"What do New Mexico dreamers dream about, sleep testers wondered. Tonight's show will reveal the truth. I'll give you a hint about the startling answer: for one thing, they dream about Mars! Stay tuned for our interview with Dr. Shing, after this message..."

Norm chuckled.

"You awake, huh?" The trucker said. "Here," he offered. He turned on the video to go with the radio. The talk-show host's face hovered down near the center of the dash. "Want to drive again? I'm tired."

**

They must be in New Mexico. The New Mexico early a.m. farm news was going now, after a couple of hours of people talking about seeing extraterrestrials. Norm was still driving, enjoying it.

In New Mexico, the farm news said, hydroponic winter wheat was ripening early, producing the fattest grains in a century according to the state Department of Agriculture. Weather stations had reported record snows in the mountain areas and record rainfall all over the state.

Around Las Cruces, it said, the cactus were fattening and going into winter bloom this year, something the state horticultural department had been unable to explain. A few dry creek beds around Las Cruces, ones that usually ran with water only after a rainstorm, had been running full for three months. The Las Cruces Livestock Association reported that a record number of mares were foaling this year.

Norm turned up the volume a little and listened harder. The next part of the story was that investigative reporters from the largest Las Cruces newspaper, *Space City Times*, had actually put together some numbers and made some maps and graphs that showed there was a focal point for all this stuff. It all seemed to be concentrated just south of Las Cruces, where there were only farms, ranches, and a campground.

"*Space City Times* reporters will not cease their investigation of these phenomena. *Space City Times* predicts that investigation may uncover a clever local hoax, and urges listeners to subscribe to the *Online Times* for more information as the investigation continues..."

Norm was getting sleepy. Better wake the guy soon.

But could these super-snoops possibly be pointing their telescopes at Ozzie's place? *Maybe they're talking about another campground. There may be more like his around there...*

Then the thought of the cactus in bloom made him think of the PII office, where the plants were overflowing their pots outside the recording studio. And of the Mars greenhouse, bristling with beans and tomatoes. And then, of the Singer.

CHAPTER TWENTY-TWO

NEAR DAWN, THE TRUCKER LEFT HIM OFF at the junction of two roads, saying, "There's a campground nearby."

"What city are we near?"

"I dunno," Sam said. "I'm not from around here. Don't have a map of this part. But they'll be glad to take your money for a night's stay, I'm sure." He grinned.

Sort of encouraging, sort of not. They shook hands.

Norm walked alongside the road, keeping to the shadows, glad for the dark moon and overcast sky. He thought of Ozzie, and stopped, shrugged off his pack, got out his knife and the sheath that went around his leg. Put it on, stuck the knife in. Considered using it for defense, and disliked the idea so much he put the pack back on and began walking twice as fast, as if he could walk away from that idea.

He strode as quietly as possible in the direction the trucker had pointed out. But there was no sign, so he thought about sleeping here, in the trees. He had no idea what dangers there might be in New Mexico forests. Did they have mountain lions? Bears? *Might have done a little homework.* He walked on, listening.

He saw a simple sign on a tree: "Campground," and an arrow. He walked faster. Now, up ahead, he saw a flicker of flame. A campfire, late as it was? He turned in on the entry road and walked past a number of tent sites, quiet and dark, toward the distant firelight. How would he get signed in? Not sure how you checked in at these places. Nobody around to check him in, it seemed.

The road seemed to meander. He wasn't sure what meander to take. The fire was no longer in view. But there was an empty tent site just off the road. He set down his pack, put his sleeping bag down on the ground pad, eased himself in, and was asleep before he could even take his sweater or windbreaker off.

**

In the night, in his dream, he heard singing that sounded like Egypt. Or Mars. When he woke, a white cat sat beside him, looking at him. Maybe dawn was breaking; were his eyes fooling him, or was the sky getting lighter? Somewhere, there was singing again, only he could tell now that he wasn't dreaming. He packed up again, shouldered the pack, and crept toward the strange music.

A stick snapped. Before he could pull out the knife,

his wrists were caught and held tightly behind him.

Norm tried to wrench away but they were strong: hands like iron on his arms. Pulling against them just twisted his back so it hurt again, badly. How stupid to get caught. He wished Ozzie were around. Or Alexis, or Freya. His breath steamed in the frosty air, and theirs did too.

Their shirts were pretty clean-looking. But they still looked like thugs. Weren't they cold? They pulled him along through the trees and between tents to a clearing, then across a fire circle with a small fire on a pile of glowing coals in the center, to a square tent that faced it. No people were around, and there was no singing now. Light showed through a slit at the tent-flap. He could smell cloves and cinnamon, coffee and garlic.

No one said anything. The flap opened and out came a guy in a white shirt, open at the throat, and loose jeans. His dark hair gleamed in the glow of the firelight. They presented Norm before the guy.

The head thug. Great.

The guy just looked at him. After about 30 seconds Norm was starting to squirm a little inside. *Is this voodoo?*

The look continued. What the hell. Norm decided he had a right to look too, then. He followed the guy's legs down to his feet, which were enclosed in simple sandals. And no socks. *Maybe his bedroom slippers. His feet must be cold. And he's pissed off because I woke him, probably.*

Something curious sat on the ground by the head thug's feet, oddly familiar. Norm's eyes focused on it. A large jug.

"Hey." Norm found the guy's face again. "Any chance your name is Malo?"

**

He was shivering by now; the October cold seemed to leak into his windbreaker from a dozen places.

"Coffee?" Malo offered.

"Sure," Norm grinned. Ozzie's friend, and a safe campground. He had really lucked out. Maybe he had some good karma from something. The other thugs—the other gypsies—melted away into the trees, maybe to ward off truly unwanted party-crashers. *I must be the desirable kind of visitor.*

"Heard about your place," he said, looking around. So this was the gypsy camp, home of the famous Malo. Malo showed him to a spot to sit near the fire. "But I never made it out here." He winced as he sat. That scuffle with Malo's fellas had not helped his back any.

Malo looked at him scientifically for a few extra seconds, then said, "Stand, please." He held out a hand. Norm took it warily and the hand pulled him back to his feet. Malo grasped his shoulders and moved them; then went behind Norm and moved his shoulders and neck; then collared him with one forearm and pushed with the flat of the other hand. Something popped.

Malo went through the tent flap to get the coffee and cream.

**

"Malo means 'bad one,' doesn't it?" Norm said. "How did you get *that* name? Get into trouble in high school?" he felt at home enough to give Malo a grin.

Malo smiled ironically, a fast flash of teeth, and handed him a cup. His fingers were brown and seamed but strong-looking. "Got the name long ago, when the Padres were here. I didn't go to their churches, so they called me the name and everyone else began to use it too. I found that it saved me some trouble because few people bothered me."

"Hey, good idea!" Norm had to laugh. They sipped companionably, and clouds of coffee-steam surrounded them in the bitter air. *Cold as Mars here!* he thought.

**

The cream was so good it made the coffee OK. Until Norm's mug was empty, they talked dreamily about little things. The next day, Norm couldn't remember them. But at first light when Malo showed him his pack sitting nearby under a tree, and his sleeping bag neatly unrolled, he had no thought left except to sleep.

And when he woke with the high sun in his eyes, his back had no pain. When he stood, sat, squatted, and bent, testing it, there was no stiffness left at all.

No one else seemed to be awake. Or were they long gone? *Have I missed breakfast?* He was ravenous by the time he returned from a trip to find the campground bathroom.

He noticed two cats at the foot of his sleeping bag. That bag was probably a cat-hair magnet. He knelt and began to roll it up. Then he saw another few in the speckly shade beyond them. When he looked around, he saw that he was surrounded by cats, sitting, standing, and arriving. *Good thing I'm bigger than they are.*

They moved closer and watched him strap his rolled bag and pad to his backpack. Then one said, [What do you know?]

"Huh?" He couldn't believe he was hearing them. It would probably always be a surprise, he decided. [Not much.] He grinned at his own humor.

The cats crowded closer, as if they all had the same idea. Norm wondered what it was, a little uneasily.

[Tell.]

He thought he'd better. Polite relations with the local cats and all. Something he knew that might interest them: [I know that this is where the famous Ahanith is supposed to be staying.]

[She is here. Your answer is right,] the front cat said. [And you must help her sing.]

"Help her? How?" Such a weird thing for the cat to demand, that he answered out loud. When they crowded closer, he added more politeness: [I'm not a singer. My singing sounds like cats fighting. You don't want me for this job!] He grinned.

The front cat sat unbudging in the pine needles before Norm's knees. Abruptly it twitched both ears, lashed its tail once and back, then curled it precisely around its feet again. The creature looked at him stubbornly. [Want you to help her sing.]

Norm looked at the cat, baffled. He looked around at them all; they all wore similar expressions: stubborn and insistent. He chortled at this absurd scene: stubborn cats giving him orders.

He found that one hundred simultaneously hissing cats had a surprising amount of impact.

[Do not be SSSTUPID!] the front cat shouted fiercely. It echoed painfully in Norm's ears and he flinched.

[Listen: want you to HELP her sing!] the cat shouted. [Think about this and do not be stupid anymore.]

All the cats turned, almost at the same moment, and walked away.

CHAPTER TWENTY-THREE

THE BRIGHT SUN WARMED his hair and felt good on his skin. Norm hefted his pack and headed for the center of the gypsy camp, and as he walked something came back to him. Sometime last night he had asked Malo:

"How come the cats keep influencing us with dreams?"

"They give you dreams?" Malo smiled, amused.

"Do they give you dreams, too?"

Malo shrugged. "Tell you something about the cats: really, they have no power to give people anything but their own dreams. They're like moons that reflect light. They could only influence you by showing you what you could already believe, or what you want to believe."

He'd have to think about that, when he had a minute.

**

She's here all right. I was beginning to wonder.

When Norm found his way back to the fire circle, he

found a couple dozen gypsies sitting around the coals of the fire and listening as Ahanith began the song Freya called—what was it?—the Growing Song or the Life Song, something like that. He recognized her voice from the day they had recorded at the PII studio.

Ahanith was at the open end of the circle; Norm could see that because it was the place the gypsy men, women, and children were watching so closely. And after a little, he could see her faintly at times as a sort of blur of white tunic and a hint of green eyes.

Her voice came from that spot. It took off pretty much like a normal song, but then it started swooping upward and did some curlicues like pea vines and then vibrated like one of those speeded-motion holos of green shoots shaking off the husks of the seeds. He was surprised at what clear pictures the music made.

Next the gypsies started in, and it was just unreal the way they could follow her. Unbelievable, in fact. *Ahanith's bizarre song sounds really good when the gypsies sing,* he thought. *Alexis' friends in the Oxbridge Chorus could take lessons from these guys.*

Norm put his pack down and sat under a tree outside the circle. Looked like the gypsies were settling in for a good long singalong and no food was coming anytime soon. He rooted around in the pack for a protein bar.

**

Maybe he should just leave. What was he hanging

around here for? Here, where the cats were bullies and there was no visible food. And Norm knew how much road was ahead of him. Alexis was waiting.

Maybe he was waiting for lunch. He hoped there would be some, and that it would be hot. The singing continued with gusto, though. Someone handed him a steaming mug of coffee and he took the hot stuff gratefully to warm his hands.

Now someone new walked into the clearing out of nowhere: a woman in tekryl hiking boots, a tunic and a cold-weather cape. A scarf was wrapped around her neck. His eyes went wide; it was Diana, Ozzie's mother.

Since there was no one else free—and besides, he was curious—Norm decided to be the welcoming committee and rounded the fire circle to where she stood.

"Norm." It was a motherly tone of voice, always kind of nice to hear.

He told, as quietly as he could, how he had arrived at this location.

She herself had come to do a daily recording of the Singer. She showed him her equipment as she set up, including a slender tripod with a multidirectional recording device on top. For "natural sound."

While she snapped pieces together he whispered to her that Ozzie and Freya and Alexis were surviving OK in England, which she knew. Also that Raker was threatening Ozzie and Freya with extradition, which she didn't. That one made her think, he could tell.

She hit the Start button on the side, held up one finger at him, and led him off into the trees away from the dozens of singers who had accumulated at the fire. She wanted to know: "Are you OK? Have enough food?"

"Yes," he lied. His stomach roared at him in protest.

"As you can see, we're still recording, still experimenting. I do this part because Ahanith and the people here are more comfortable with me. I'm familiar." She told him that by the time when Ozzie "disappeared," (she seemed to use the word deliberately) she had done tests enough to show that it would take more of the same sound to make a difference in the earth's crust fast enough. But what was "more"?

"You've heard: we're finding that more singers seem to make a greater difference than just more loudness or more electronic power does," she said. "We're asking this gypsy group to add more singers daily to test the numbers involved. They're up to almost 120 now."

She talked about the evidence that the music affected things closest to it first. "Have you heard the stories about the record-breaking crops and livestock within a 30-mile radius of this place?"

"Yes." He tried to sound scientific instead of hungry and cold.

So she was looking at ways to get the sound nearer to the worst areas of disruption to help them faster. Maybe Europe because it was nearer to the shakiest area:

Iceland and the Mid-Atlantic rift. And other volcanic areas, like Central America and the west coast of the U.S.

She was talking about transmitting recordings in those areas? He was having a little trouble putting it all together well, trying to think through his calorie-starved brain, but in the absence of clear thinking he thought of karma and offered to go fetch her some coffee. When she accepted, he went after it and decided he could get away with pouring more for himself, too. With lots of that great cream in both. You could almost live on that stuff.

It was while he was returning to her, treading carefully in the leaves, sticks and pinyon needles to keep two full-to-the-brim mugs from slopping over, that the idea came to him.

CHAPTER TWENTY-FOUR

"SO WE COULD MAKE LOTS OF COPIES of Diana's recordings, see, and get them to the gypsy camps across North America. Get them to sing, too, along with the recording." Norm held out his hand for the old player Malo had dug out of his tent.

"Here. She—Diana—gave me a data sliver with today's recording on it. Also this sliver, a recording of just Ahanith in the studio. You can copy them like this, just plug in a new sliver, plug in the things to copy, and hit this button. One's copied. Now both are copied. And now the copy plays, like this."

Malo listened, half-smiling, intrigued. Norm listened too, with satisfaction: there was the whole package of sounds coming out of the player, voices clear and pretty strong: Ahanith and the group of singers here.

"And they could just play them on a cheap player like this one, see?" Once again Norm hit the Play button on the thin metal slab. There was the music again.

Malo sat back and sipped at his mug of stew, thinking.

Well, Norm would have called it thinking at first, but it seemed like more than that. You could just see things building there in the air, players and places, singers and songs, as he considered all the parts of what might be done. It was like Malo had his own holo-design software going in the air.

Malo nodded. His images vanished. "We should do this. Come." He led the way from the early dark outside through the flap into the bright hot light of his camp stove inside the tent. And the smells of cinnamon, cloves, and other things. Like the ones in Egypt, the smells in here were mysteries to Norm.

Within seconds, it seemed, Malo had water heating, cushions thrown down on the floor rug, paper and pens.

Norm felt like a genius consultant. Clearly they were about to make the battle plan for his idea: how to propagate Ahanith's sound to the other gypsy camps in North America.

**

When they had lists and diagrams all around them on the low table and floor, Malo pointed to one list:

"Here we have friends at other campgrounds all over the Rockies," he said. "And here we have runners who can take them to these places." He scribbled some more. "And transport like the truck you rode to California can take them further for other friends to relay northward on the coast, to these places: Los Angeles, San Francisco, Portland, Seattle, Vancouver...And they can make copies to pass on? Good ones?"

"Yes, pretty good ones. They get worse the more you make copies of copies."

Malo frowned, thinking. "Then we need slivers that they can copy, to go with each. For them to make more good copies. And we need 100 for you to take as you travel to the east..."

Norm began to add up the number of everything that would be needed for all locations. This was getting big fast. *Who pays for this? And I'm going to carry 100 of these? Good luck.*

In the end, while Norm sat and rummaged in his mind for any more equipment they would need to use, Malo sat and added figures, divided, showed percentages and interest rates and other fancy math. "Want to use my calculator?" Norm asked.

But by then Malo was finished. He grunted and rose to dig through a wardrobe. Norm searched the bottom of his mug for any more meat or vegetables left from his stew. This had been his third refill. Probably this was it for the day, but he wouldn't mind more.

Malo turned and held a flash of red up between a thick thumb and his forefinger: "Very old ruby," he said. A couple of dark forelocks had fallen over one eye and he brushed them back with his fingers.

Norm's eyes grew big at the stone. "Fat one," he nodded. "Must be worth a lot of cash!"

"For the players, tomorrow. At the price you said, we should be able to get 200 or more with this. Now, a list of stops for you to make, and instructions to go with copies and players at each one. You can drop ten at Albuquerque, ten at Amarillo, ten at Oklahoma City, ten at Tulsa, ten to the Cherokees (like the gypsies, they are good singers too) and ten at Little Rock, ten to… I will decide. You can give these to them, show them how, and they will run them north and south to the campgrounds in both directions, one player for each…"

"But will they believe me?"

"Probably not. So, I will send letters from me to go with them. My letters will say you have free stay and food at any gypsy campground to the Atlantic Ocean in trade for delivery of the letters and players.

"Most of the ones on this list will have a way to pay for copies and more players. I'll give you some small gems for the ones who don't…"

**

Norm looked up at Malo from the messages on his phone. The little stove was out, and only a candle burned.

"They are some pretty nasty guys, the ones who are after Ozzie and Freya. It worries them. Kind of worries me."

Malo nodded. "Worry doesn't help though..." And after a little while he said, "Who do you believe in more, you or them?"

"Huh? Me. Of course."

"Good to hear." His teeth were very white. "Then you should never act like you believe in them more than you."

**

Because Malo's bargaining skill was something Ozzie always talked about, next morning Norm wanted to go with Malo to see him bargain at the electronics store in the city. But Malo decided he should stay and copy slivers, test and label them, group them into sets. "Important job," Malo said. "You should do it alone without interruptions. String will keep you company," he added.

String must be the black and gray tabby cat that was looking at him so intensely.

[Yes,] she said.

He mused: *Malo might be worried about the danger if someone recognizes me.*

[Yes,] she said.

Norm looked at her a while, then sighed and got started.

**

By midafternoon he had the job done easily, and

tested, and everything ready. He wasn't looking forward to carrying the 100 cheap players. Still, they wouldn't be too heavy. And at least his back wasn't hurting. *Hope the magic or whatever he did lasts.* How would he fit all the new junk in his pack? No idea.

He looked around in Malo's tent, bored. Nothing much to do here.

Then he had another idea. He'd make another few sets of master slivers, in case they came in handy in England. And another thing…

"Want to come help, String?" He decided to treat the cat the way he treated his little brothers when he had to babysit them: keep them on his side so they weren't stirring up trouble and tattling on him. Maybe String would keep that gang of cat-bullies from slashing him to ribbons, too.

**

String had grudgingly agreed to come with him to the student section of the campground, to be his "lookout." Right now she was outside guarding while he did business in a student tent with two guys who had loads of camping gear. Norm traded most of his clothes, his remaining food, except some protein bars, and one of his game disks for a used holophone in good condition. For Malo to stay in touch with him and Freya, Ozzie, and Alexis.

The fancy backpacker radio in the tent was playing

local news, about that Space City newspaper and its hunt for the source of the "Las Cruces Hoax." So Norm decided to get the inside story. He used the privilege of a curious traveler to ask what they knew about it.

One said, "The investigative news team is narrowing in on where it's coming from, whatever it is. Looks like it's someone in this campground, unless it's Reed, the rich guy who owns all this land."

Norm thought of Ozzie's dad, and how rich he didn't seem to be.

"So what are they gonna do now?" he asked, grinning gamely.

"They're planning to arrive any time now! As soon as they have the right coordinates for the source of the hoax located, they'll be here with cameras and recorders, to expose the whole thing. Should be pretty interesting."

Norm thanked them, with handshakes all around, and took off for the gypsy part of the camp. He went at a run, now that the pack was lighter.

CHAPTER TWENTY-FIVE

NORM RAN UP TO MALO at the fire circle and shrugged off his pack. The singing faded out. Malo had a "we've been wondering where you were" look.

"Malo? I'd better talk to you—"

"You must go soon. And you must take Ahanith with you," Malo said. "Too dangerous for her here."

"How did you know?" Norm said.

"Holoscreen at electronics store."

The players were bundled into a pack-sized brick. Norm could see that each group of ten players and ten slivers had been separately bundled into a layer with a letter, and wrapped with air-wrap. Malo oversaw his loading them into the bottom of his pack, then watched as he put his few remaining clothes back inside. He gave Norm a tiny envelope of small rubies "for expenses," which Norm stuffed deep into the cargo pocket with the cash.

"Your pack is too full for the pot to go in. It will draw attention. Not safe." Malo disappeared into his tent and returned with a large, very worn goatskin, flat and sand-colored, and he invited the Singer in, then held it in his arms.

"One more song," he said. The guitarist started up again. For the first time, Norm heard Malo sing too.

Look at him making love-eyes at her. An ancient guy, oldest person on earth I'll bet, and a female with no body who's in a goatskin bag at the moment. Wowee. Pretty way out. Last year Dad looked at Mom that way at Aunt Flora's, and she blushed and whispered something to him to make him stop...

Malo held the old leather thing against his forehead now, eyes closed, as the last of the song wound into the air from the goatskin and the circle at the fire, surrounding him.

When the song stopped, Malo knotted the thongs of the Singer's goatskin flask tightly to the top of Norm's stuffed backpack. "Take care of her," he said.

"Take care of him," he said to Ahanith. His eyes were soft with amusement and woe.

Right now, the entire human content of the gypsy camp seemed to be arriving at this spot. Norm had not yet seen them all; there were a lot of them.

Within a minute, women with children had arrived and slipped quietly into tents around the fire. Some of the biggest thugs (Norm thought it with some respect, now) took up stances at the flaps of these tents.

Two men stoked the fire to a nice crackling blaze while many other men and women seated themselves, and the guitarist re-tuned the guitar, beginning to sing an old ballad.

"Malo? What's this?" Norm asked.

"Last songs for Ahanith," he said aloud. Then he leaned toward Norm and added, "When the reporters come they will see this and report it as the 'hoax.' We have a silly story to tell them about singing for peace so they can put it in their papers. They will tell a story that makes us seem like fools, then they will be happy and go away after a while."

"But you can just *make* them leave. This is private land, right?"

"Sure, but that's a lot of work. Easier to just sing some songs and make them go away that way."

As fast as people were arriving at the fire circle, the cats seemed to be packing up and moving out. Norm could hardly focus on them, there were so many; they traveled away from the fire circle in all directions like schools of fish, gliding along the shadowy forest floor through the trees.

One black and gray tabby arrived at their feet, though, and stopped.

Malo's eyes went softer and a little more woeful. "String has decided to travel with you, Norm. She is wise and she knows why she is doing this. Take care of her."

An old truck, older than Ozzie's, stopped out on the road nearby and sat with its motor running. A dark-haired boy was the driver. "Qualen," Malo pointed. "You will ride with him until it's time for you to go on alone," Malo said. "He is 15. Take care of him, too."

The same age as his sister. Malo nodded soberly. Norm felt burdened. He sure had a lot of people—or whatever—to take care of. They walked to the passenger door of the truck.

"Wait. Wear your disguise. Now—quick!" Malo said. Norm struggled to open the pack up again and scrabbled hastily inside, his fingers searching for the goatee and mustache and hat. On the way down to the disguise his fingers also felt the extra phone.

"Oh. For you, Malo. To call me and Ahanith. The number's in there."

The man's teeth flashed joyously. Norm slapped the mustache and hat on, stuffed the goatee into his pocket.

"OK, quick. Get in." Malo had his pack closed and in the back of the truck with the bags of trash before the others shut the door behind him and String. "Go now, Qualen. Be at the turn before..." Qualen gunned the engine and sent dust soaring behind them.

As they jounced down the road he said to Norm, "They're coming. Hat on, get down."

I sure am getting a lot of orders from these guys here.

Qualen must have made it to the turn on schedule, because he pulled aside and slowed, and he allowed

Norm to look back—just in time to see the last of a caravan of news-trucks rattling importantly down the campground road.

**

It was cold on this dark highway. Good thing he was ready for it: he had brought a windbreaker and some sweaters. That was all he really needed at Berkford. Could Ahanith feel the cold?

[Yes,] he felt her say. He was pretty proud of himself for hearing. The truck driver, who was asleep on the passenger side, wasn't listening.

[Uh, want me to have String sleep on top of your goatskin?] String woke and sneezed. Or snorted, maybe.

[Thank you. No. It was cold on Mars too. Cold is cold to me, but it can't harm me.] So Norm snagged his second sweater and dropped it over String, who climbed back out of it and lay on top. *Cold-weather cat. Great, now my sweater will be a hair-catcher, too.* String opened her eyes and gazed at him tolerantly as he drove another turn of the winding, hilly road.

Well, it had been a fast hop to Socorro with Qualen, and a fast drop of the slivers and players with the gypsies at the little campground there. Soon, Qualen said, Malo's messengers would be traveling north through Albuquerque to Montana and Wyoming, and south to Mexico and beyond, with their players and sound-slivers, but this way the Socorro campground could get started early. They were not fast people, he explained.

"Change of plan," Qualen announced. "You go quickly from here, east to Roswell on the back roads." And he waited, parked in the shadows, till Norm caught a ride. From there, Norm was relieved to know, the fella was headed back homeward to safety near Malo. He was too young to have to deal with being attacked on the road.

He was also probably too smart to, Norm reflected. He texted Malo, although he might not know enough about phones to look for a message: "Qualen is on his way back."

Ahead was Roswell, New Mexico, famous for being a pit-stop once for some unfortunate extraterrestrials. They were almost there.

The sleeping driver had been so grateful Norm was willing to drive that he let Norm name his drop-off place. It was the Roswell campground. The fella might be surprised to find himself and his truck there when he woke.

**

The Roswell gypsies woke him for breakfast; he had slept by the fire circle this time to make sure he didn't miss out. They seemed to be expecting him. Nice people, good food. One of the little kids was interested in how he got freckles. They didn't seem to have any freckles among them.

The gypsy group here messed around with the players till they got the use of them down easily, and then they sent someone to get him a ride while they had a huddle about where to take the nine extra players, north and south of here.

And then he was on the road to Portales with one of the messengers; there he could catch a ride to the campground in Amarillo, Texas.

**

In Amarillo Norm texted Malo: "String and Ahanith are OK, and the rides are OK. Watching out for weird truck drivers. Dropped stuff in Roswell and Amarillo. Still have newshounds there?"

Malo sent a voice answer: "The news people left the same day. We cooked really good stew, ate it and didn't offer any. They got hungry, went home to eat and didn't come back. The gypsies are still singing and recording. Some students have come to sing with us because the newspapers say it's a peace song. The veterans of the Wars haven't visited yet."

**

The quaking gave Norm the creeps. It started in Amarillo as something you felt now and then. But once he was on the road through Oklahoma, there was no break from the volcano at all. No wonder Freya was so hyper. He felt a little foolish to have been so tuned-out about the way things were for her.

Yeah, this is really disgusting. No one would want to live with ground that shakes all the time.

[When the ground shook on Mars...] Ahanith said, [...it was a bad time] she finished lamely.

[When was that?] he said. But she had drifted off again. She was sleeping a lot on this trip so far, as if the daily songathons in New Mexico had been tiring. Probably they were.

Oklahoma City had a sort of impoverished-looking campground, but the head guy welcomed him. From deep inside his pack, Norm pulled off the topmost slab of players and slivers, wrapped together in the thin layer of tekryl air-wrap, to give to the man first thing. Lighten his load. The nice thing about these deliveries was that his pack got lighter as he went.

It was midafternoon and they looked hungrier than he was. After showing them how to use the players, he told them he just needed water and a ride out to the highway again, and left them ten of his protein bars. He'd probably have enough anyway.

CHAPTER TWENTY-SIX

"WE DON'T TAKE UNMARRIED couples here," the hostel manager said. Freya thought he must not like their unmarried-sounding fake names, the ones they had used to get student monthly travel passes in London: Francie Bell and Arthur Rimes.

Freya and Ozzie walked back out the entry door and away. It was several miles' walk to the next hostel in this city, Birmingham. Freya was weary from the weight of the pack.

"Alexis said it's getting traditional in England again, Ozzie. Maybe more so as we go further away from London." They were both more tired than they needed to be; probably from too much worrying.

At the next hostel they entered as Francie and Arthur Rimes. For that occasion she wrapped string around his ring and pinched the thick band carefully between the two adjacent fingers to make it look right long enough for them to check in.

IN THE RING

Ozzie's cheeks had reddened with embarrassment when she asked to borrow the ring so they could pretend to be married. It was a delicate subject, she guessed. But they needed to room together for safety. And they weren't walking all the way to a third hostel tonight if she could help it.

**

Ozzie just couldn't resist it: After they locked their packs in the "married dormitory," he dragged her out with him to look around Birmingham in disguise. They walked into coffee shops here and there, browsing through the menus, and Ozzie practiced his British accent by asking the servers: "What do you think of those volcanoes?"

"Yeah, bad show for the Icelanders," was the most usual answer.

"Some of the lava leaks are also in America, and Japan. What if they come here?" Freya would ask, trying to sound like a Brit named Francie.

"Oh, but that's not likely, is it?"

If anyone asked where they were from, they decided they didn't want anything and left. And went to a park bench somewhere to practice again.

**

The next day, traveling by bus northward through England, Ozzie picked up more worldweb reviews on

Ilse and Doug Reed's latest performances. She and Ozzie read them.

"Maybe we should just go like mad, straight to Reykjavik," he said.

"Why? We can't stay with our parents," she reminded him. "Hostels here or hostels there: what's the rush?" Maybe one of these British libraries would have the information they needed. For now, she leaned back into the worn bus seat and poked at her phone some more to continue the worldweb research she was doing.

**

Early morning seemed like the safest time for them to travel.

Now they rested for a minute under the pre-dawn lamplight on a bus stop bench in Manchester. "Newsfeed," Freya said. "Oh, Ozzie. He really did it: Raker made his rotten extradition stick. Yeah, they passed it! We officially have a bounty on us."

"Damn."

"We'd better use these disguises round the clock and the fake names all the time. OK, Arthur?" She made a face at him. He grinned.

**

There was frost on the ground as they got off the bus at the Edinburgh station. Freya shivered. She had all her warm layers on: the sweaters and her hooded sweatshirt—which were all she had carried with her to New Mexico in June—and the things Alexis had given her.

At Central Library, Edinburgh, they spent two days researching more about qualities of sound and the ways sound affected things like glass, dirt, rock, the earth's crust. They pushed farther, understanding better. She was encouraged. But Ozzie complained, "It still doesn't seem like we're getting where we need to go."

And today, their third day here, they saw signs that they had been located. The signs were familiar: an unusual number of visits from the librarian. Then a man who passed by their table and eyed them both, disguises and all.

Ozzie had scoped out the floor plan of this library, and they had been through it before, at the Manchester library, so they knew the drill:

They rose, shut their books and put them onto the nearest automated shelving cart. Each of them went into a bathroom. There they straightened up their disguises and left separately through the back exit.

They walked by different routes to the bus station and used their keys to get the packs that were stored in separate little rental lockers there.

They sat in separate booths in the station restaurant eating lunches that took hours, pretending to be arguing politics with someone on their phones every time the waitress came by, until their bus was called.

They boarded separately northward, using their student monthly passes.

Later, after many of the passengers were sleeping, they relocated: Freya met Ozzie in an empty rear seat and they talked quietly. He had the maps. She had a bad feeling but she didn't say it.

"Not much room north of here to get lost in," Ozzie said. They looked over the map: Northern Scotland had lots of space and hills and lochs but not so many places and people to hide among.

"Sorry, Ozzie, I was wrong; think we should have left England a few days ago."

"No, this is OK. The last stop on this line is Thurso ..." It was a harbor town at the very north of the mainland. From there, sometimes you could catch a boat to Iceland. Her eyes went with his as he looked out into the darkness. Thurso would be another few hours beyond the hills that sat here sleeping in the cold and dark—dark except where a lonely house or farm light shone onto the snow.

To encourage each other, they talked about the next leg of their journey till they both drowsed, and then fell deeply asleep.

"End of the line!" the driver called out.

When her eyes flew open she saw Ozzie's startled face near hers. *Damn! We should have been ready.* Dawn already lit the sky. They hurried to gather up their packs, but they were the last ones to get off the bus. Together.

And they had been spotted, she knew it. Freya turned just a little: a pair of men began to follow them on foot, walking quickly.

She nudged Ozzie, turned a corner, and they ran. They dodged down alleyways and up side-streets, keeping the gleam of the water in view as much as they could.

She was hungry. He must be too. And she felt out of shape: getting winded. Too many weeks in the library. Ozzie panted, "Damn. *Can't* go to the boat docks: they'll expect that!"

"There! Come on, Ozzie!"

He ran after her, on faith alone she knew. *I hope I'm right.*

It was a fishing boat—and her eyes hadn't fooled her: it *was* an Icelandic name on the bow—moored at the wharf. She slid to a halt on the gravelly cement, swung her pack onto the deck and jumped on, with Ozzie right behind.

Freya said breathlessly, as the boat owner appeared from the wheelhouse, "*Við erum nýju áhöfnin.*" ("We're your new crew.")

His mouth opened in protest. His beard was sprinkled with a scattering of white. He reached back into the wheelhouse, the roofed shelter that was like the driver's compartment of the boat, and cut the sputtering motor.

Við munum vinna ókeypis ef þú tekur okkur! ("We'll

work for free if you take us with you!") she added.

A plain black car pulled up alongside the wharf. The windows were tinted. The doors opened.

Vilt þú? ("Will you?") Freya begged.

The boat-owner sized her and Ozzie up for half a second more, took a fast sideward look at the men getting out of the car, and cast off, shoving hard at a tall piling to quickly put distance between the boat and the wharf. When the men reached the wharf and shouted at him he shrugged helplessly; could he help it if the boat was already launched?

Freya, who had started the engine again, watched anxiously from a shadowy corner through the port-facing wheelhouse window. The boat churned water.

By the time the owner reached them again his new crew was already at work there inside the wheelhouse, where Freya was showing Ozzie all she knew about the boat: the basic instruments. She hoped their captain didn't expect more.

**

Ozzie slapped hands with her, quietly. "Good one, Freya!" he breathed. Still panting, they did what the guy had just ordered. Freya reported that his name was Völundur and he seemed to have a good-sized list of things that had to be done fast. She translated.

Völundur's cheeks were red and polished by the cold air, which smelled salty and wild. And the air was very cold; colder than they were dressed for. The man sent Freya below to get jackets for them both, which she toted back up the ladder. Ozzie took his gratefully.

It wasn't a large boat and it had to be at least 50 years old. In fact, it seemed pretty small to be traveling out on the open ocean. Now that they were headed in that direction, Ozzie said so to Freya.

She nodded. Her green eyes shot sparks. "But you will be surprised. To be safe, just do what he orders."

When they were underway and leaving land behind, Völundur sent them back down the ladder with their packs. But not before he gave Freya some very deliberate instructions in Icelandic.

"What was that about?" Ozzie asked her as she lowered her pack down the ladder to him.

She took him to the berthing area and showed him his tiny bunk, below Völun's, and hers, opposite. "Remember Norm's Bunk Rules on the Mars flight? No changing beds, one person on a bed, and all that? Völundur just told me his version of the Bunk Rules," she said.

They hauled nets up to the deck, cleaned the kitchen and the hold (which stank of fish) and scrubbed the decks until he was sweating in his heavy jacket. Maybe the guy was trying to get all he could out of them while he had them, but it felt good. Ozzie felt like he was scrubbing away the stuff he hated most: sitting quietly, being polite and cautious with people who were not direct, riding on trains and buses. He scrubbed still harder, recalling the weeks in London.

"Does he know about the extradition?" he muttered to her.

"No, and I'm not telling him. Better not to know." Freya's eyes shot sparks at him again. "Pay attention, now, Ozzie. No people overboard, here," she said and winked at him.

When did she get so sassy? He grinned. She must be in her element.

His stomach was complaining now: hunger. But again the guy called out something that sounded like an order, not a lunch break.

Ozzie looked at Freya. She ran to Völun's side, so he followed and mimicked her as she worked with the fisherman to secure a net at the back of the boat—ready to be put into the water, he guessed. Sometimes the boat rose up and slid down the waves at angles Ozzie didn't want to think about too much. Völundur didn't seem to be worried, though.

Next he taught Ozzie to pilot the boat; where things could be found in the wheelhouse, how to turn the engine on and off, how to steer. Not complicated, but it seemed as crazy as learning to fly a toy rocket through the depths of space. Close up, the vast ocean was overwhelming.

Ozzie practiced under the fisherman's eye. Up the wave, down the wave. *This is out-of-control crazy.* But at least the guy was standing by in case something ridiculous happened.

He put Ozzie onto a course through the waves and told him to hold it steady while he went below. Ozzie was astonished. The waves ahead of the prow were huge. "What if Freya helps too?" he said. She translated.

No. Völundur shook his head and elaborated in Icelandic.

"Wants me to learn the kitchen," Freya rolled her eyes. "Food prep." She talked with Völundur again, probably pleading for clemency for Ozzie: don't make that greenhorn sail the boat alone. Or maybe, What makes you think I'm a cook and he's good at running boats?

Völun grinned and pointed deliberately at Ozzie, put his thumb up cheerfully, and went below. Freya took a deep breath and followed. Ozzie gripped the steering with both hands and swore prayerfully. Up the wave, down the wave. Up the wave, down the wave.

Five hours later (his phone clock said only ten minutes) when Völundur returned with Freya, Ozzie was having fun. Völundur took the wheel. "Good," he rated Ozzie's performance in English. He didn't seem to look at the instruments; just sailed ahead as if he did it by feel. Maybe he really did it that way.

He gave the wheel to Freya next, without instructions. Ozzie's hands were itching for it by the time she handed it back.

**

"It's a lot like sleeping on the space cruiser, isn't it, Ozzie?"

Well, sort of. If a space cruiser rocked back and forth and up and down on waves. At least neither of them was the type to get motion-sickness, he thought groggily.

It was still dark here below, and cold. It seemed like he had spent half the night getting warm, and now Völundur waked them? The urge to roll over and ignore everything was huge. So he dragged his legs from under the blankets and put his bare feet on the ice-cold floor to wake himself up. That did it.

"Time to catch fish," Freya chuckled at him. He raked his fingers through his hair. She was already dressed and pulling on her jacket. "He says we're past the Scottish fishing limit, officially in open waters now." She disappeared and he heard her clanking things together in the kitchen. Mercifully, she might be making something hot to drink.

At least his boots would warm his feet fast. He pulled them on over some clean jeans and socks and stumbled toward the "head," the tiny bathroom near the kitchen, bumping into things as the ship toyed with his balance.

He couldn't get phone reception now, not even the time or date. Freya couldn't either. He clambered up the ladder to the deck, where a chilly-looking sunrise seemed to be happening. Their boat chugged along, up waves and down waves, and Völundur called something cheery in Icelandic at Ozzie—it had to be something like "Hey, Sleeping Beauty!"—and motioned for him to come help. Once they had the net lowered and dragging behind the boat, the fish would boil up into it as they sailed. Up the waves, down the waves.

Völundur assigned Ozzie to the wheelhouse. Ozzie grinned. *Now this is worth getting up for.*

"Freya!" Völundur had her up the ladder in 10 seconds to help him pull in the net. Somehow they did it together, while the fisherman had Ozzie hold the wheel steady; they poured the fish into a large tub with drain holes that formed the back of the boat. But then, Ozzie thought, maybe this guy was actually able to do it all alone if he had to.

Under orders they ate some food Freya heated: big bowls of thick stew from a freezer compartment. "All I did was warm it up," she said. It was good stuff though. As soon as they had put the bowls down into the galley sink, they were to get on deck and take care of the fish. "Oh boy, Ozzie, hope you like fish," she said.

Not this much fish.

But at the moment this was a real improvement over libraries and hiding.

CHAPTER TWENTY-SEVEN

NORM'S EYES OPENED. Memphis West: a big red sign glowed diagonally in the window. He almost fell asleep again but then he sat up abruptly in the truck cab, looking out. They were parked at an all-night diner. Roger must have stopped in for a snack.

Just then the driver-side door jerked open. String yowled angrily at the intrusion. Roger, the driver, stood there horrified, his hands full of a pair of hot coffees and under attack by the cat.

"Yo, wake up. We have to go. Someone's called the cops on us."

"For what?" Norm pulled String off the driver's seat. But every time he got her claws unstuck from the places where they were lodged, they stuck in new places. "Shhhhhh-sh-sh," he said when he got her to the center of the seat, as if she were a human baby. That's what the aunts did with fussy babies. Not really working here.

"Dunno." The guy was buckled in by now, starting up the engine. "You in trouble? When I told some of the other drivers about you, next thing I knew the manager's wife told me I'd better get out because he just called the cops."

"About me???" Norm was waking up fast. "So maybe I *am* in trouble. Let's go!"

**

Although it was hard for Roger to believe that the Memphis campground would be safe, Norm knew it would be. Besides, he had to stop there. No choice; he had players and slivers to deliver. The gypsies entertained the surprised driver with hot food and some songs while Norm exited with his pack "to get OK to use the restroom" and got out another slab of ten wrapped-together players and slivers. This was the sixth; the last one had been near Little Rock. There was much more room in his pack these days.

The head-person waited for him nearby. She was a woman who looked like she could be Malo's sister. She accepted the slab and Malo's letter, read it over, and invited Norm inside a tent. Then she cautioned him: "Someone is following—looking for you or your friends, three of them? They were here many weeks ago.

"Now you must go another way. Help is coming. Your truck driver will be fed and returned to the road, but only after you are gone. He'll think you are sleeping here for the night."

She hesitated, then asked: "Maybe the Singer is with you?"

Norm knew his face must have shown his surprise. Before he could decide what Malo would want him to do about this, he felt the Singer awake and listening.

The woman's face softened and her eyes became brilliant. "Thank you," she said.

Thanks for what? No, she was thanking Ahanith, he concluded. *For what?*

She left him and his pack in the tent with the slab, ordering him to be quiet. But the order was amazingly respectful. While he stood looking around, peering curiously at her strange old furniture and breathing the cinnamon-and-cloves smell, String entered soundlessly and gazed at him.

[Norm?] Ahanith nudged him, like a mother reminding a kid about manners.

Norm looked at String. He must have missed something. He tried a little Ozzie-like politeness: [Something you want to say?]

[Been saying it. I see you are listening now.] String sneezed again, or possibly it was a snort. [You must keep me beside you at all times, from now until I tell you.]

He was puzzled. *Even in the bathroom?* But he knew better than to quibble.

**

The water whispered by them. His pack might be getting wet on the bottom, but String sat on top of it, staying dry, licking her paws in the moonlight. The last sheet of water that had slipped across the surface of their raft was pretty deep. *Lucky that the players and data slivers are wrapped and waterproof. Too bad my clothes aren't.*

He was helping to pole the rigid tekryl raft upriver, which meant that sometimes they hit a wavelet and the water surged over his feet, oozing around his pack. But this was fun. He was doing a great job of poling the raft forward and keeping his balance, while his companion poled at the front edge to steer. Like being Huck Finn and Jim on the Mighty Mississipp.

The raft ground onto some pebbles and gravel and he stumbled off. It hadn't been a long voyage, but it seemed to be over.

The gypsy guy, about his age, gave him instructions and asked him to repeat them.

Do they think I'm an idiot? It was possible.

But following the instructions exactly, he found himself safely up the riverbank and standing at the roadside at dawn, his hand out for a ride on a side road that would take him back to 40 eventually. After whoever-it-was, the people who were after him, lost track of him, he hoped.

A big truck stopped with a squeal of brakes. Norm opened the door. "Going to Nashville? Yes, just me and my cat," he said. "She's very polite. I can drive truck and I have a little food so you don't have to..."

That was the shortest interview I've ever had from a trucker, he thought as he climbed in. *I must be getting good at this.*

String hissed at him.

"I thought you said that cat was polite," the trucker said.

**

String kept digging her claws into his leg. He was used to the little brothers, so not much bothered him. But really? "Stop it, String," he said.

She dug her claws in again. Norm snapped out of the game on his phone. He looked at her. [What?!]

[This driver is cheating! Ahanith tried to tell you. Pay attention. Passed the campground, taking us south.]

Oh no.

"Hey, buddy, need a bathroom quick. Just had an accident, ugh. Food poisoning, I think. Stop here, huh?" Norm made disgusting pre-vomiting sounds, and then belched ominously too, just for effect. "Think I'm gonna heave. I'm afraid I'm messing your seat..." Norm made his belly swell, then created more disgusting noises. The guy pulled off the exit hastily.

"Gonna need to change," Norm groaned again and doubled over a little.

"Throw out anything that stinks," the trucker

ordered. Norm grabbed his pack and ran, limping and stumbling theatrically, into the gas station store.

String ran after him.

Just inside the entrance he spied a little map on the wall: "You Are Here," it said. He shot a fast phone scan of it. In the bathroom, with the door locked, Norm sent it to Malo with a message:

Need help from the guys at the Nashville campground. Don't have their number. Can they pick me up at this location any time today?

He took as little time as possible in the restroom and slipped out a delivery door at the back of the building, into some thick woods. He hoped Malo would get his message sometime soon.

He heard someone yelling in his direction, so he walked deeper and deeper into the woods. He kept the trunks of trees between him and the gas station. String stayed near him, in the shadows.

**

The days and nights blurred for him: Nashville to somewhere in Indiana, and onward from there to a campground in Ohio... without thinking, Norm thought to Ahanith: [You're probably bored with all this truck stuff.]

He was startled when she answered. [No, interesting... Drove a ???? to take the last people on my planet to the mountain shelters.]

[Drove a ????]

[Yes.]

He left it at that, but then a picture of the thing bloomed in his mind: a startling vehicle, huge, with lighted disk-like wheels and some kind of weird suspension that let the carriage rock between them. Designed for travel over rough terrain, he guessed.

Somewhere early in Ohio Ahanith had begun to sing him to sleep when he was too wired-up from driving. He could only barely hear her himself, and he was prepared to tell the truck driver of the moment, if one of them heard her, that she was just an ear-bud he had playing in his ear. But no one besides him ever seemed to hear her. Why Ahanith decided to sing to him, he didn't really know. Maybe to stay in practice.

Anyway, she seemed to like him. He tried to make sure she was comfortable, although it wasn't clear how he could do much about that. He asked her why she needed the goatskin; a pot had a little room, maybe, but wasn't it cramped in that leather bottle?

[Liked the jar better,] she admitted. [I'm tired so I feel safer having somewhere to be.]

Finally he settled on telling her what he knew about: funny stories from Dingo's Pizza and intriguing robotic ideas. She was equally interested in both. In fact, she was the best listener he had ever run into, once you got the hang of talking to her.

Here they were now at the Baltimore campground, last stop. Norm knew that Alexis and the others had been barred from airports by Raker's ridiculous search for them. It would be the same for him; he would be discovered and questioned (or maybe worse) if he showed up at any airport. He imagined himself fleeing down an airport concourse with several killer-robot dragonflies homing in on him.

But he had been so intent on getting the slivers and players given out, and fast, that until he unloaded the last slab of them at the Baltimore campground he just didn't look much into the problem of how to get across the Atlantic.

When he called and mentioned that to Alexis she had a cow, completely. "Here I've been worrying about you, not even knowing if you were safe half the time, and I could have been helping you with a solution!"

"Well, I didn't have time to be helped," he said carelessly. Then he thought about karma and all that stuff and said, "Hey, Alexis, thanks. You are a peach. Really. And I do need you to help me now if you can."

While she was doing research by phone, riding on the high-speed train homeward from Oxbridge to London, Norm hid from the campground kids, who wanted him to play softball. He tiptoed off into the woods and sat behind a tree. He turned his phone sound low and blew up pursuing aliens until the sun went down and he decided it was safe to show up again.

Alexis called back when he was waiting in the dark behind the campground bathhouse. String sat beside him against the cold concrete wall, sweeping her tail restlessly back and forth until his phone signaled the incoming call.

"I've been trying to reach you! What did you do to your phone? Quick, Norm! Get to Baltimore harbor!"

He said he had a few little jewels, hidden in a place he wasn't mentioning for reasons of security, but not enough spending cash for the ground transportation.

"I'll hire you a shared ride, then."

**

He caught the shared ride by gliding out to the roadside from tree to tree, and standing just outside the pool of light from the campground office till the hoverbus swung into view. At the last minute he thought about String getting onto the bus with no carrier, and stuffed the cat unceremoniously into his windbreaker.

As he rode in the bus Norm decided he wasn't taking good enough care of Ahanith and String. When he asked Ahanith, she woke and said that he was doing fine. String just sniffed.

They arrived at the harbor, and Norm hastened to locate the right boat-slip among the hundreds there. Tall pleasure-yachts floated side by side with a bobbing shantytown of littler boats whose owners lived on board. It was like a campground on the water.

He had received Alexis' phone instructions to get there and get onto a boat occupied by year-round residents, friends of someone. They didn't seem to mind, so they must have known about this plan when he stepped onto their sailboat past the laundry on the rail with the pack on his back and a cat-shaped belly sticking out.

He had no time to think about it, or even chat with them. A local-transport air-truck immediately came in for a landing right behind him, on the dock beside the boat. The driver of the thing motioned frantically for Norm to get in, and after Norm had a short conference with String, who was pleased to be consulted, they climbed in very fast. String even submitted to being incarcerated in a pet-carry box as they went.

This transport truck must be taking him to his next rendezvous, Norm thought. Just like in the old spy flat-films. *I know what's next.*

But no: instead they went up a hundred feet or so, taxied out over the water like they were a sightseeing tour of Baltimore, and then suddenly hit warp-speed and everything disappeared for about 5 minutes.

Norm looked around anxiously for some scenery, but saw no sign of any, and then, just about the time when he felt he'd better ask the pilot some pointed questions, fwoomp, like they were landing in water, they seemed to hit the atmosphere again and slow down as if a parachute had just opened.

Which he could now see it had actually just done.

And there were Big Ben and The Houses of Parliament and all those things. They drifted downward, rocking gently under the chute. If this wasn't a simulation with a tourist vid tacked on at the end, he had just arrived in London.

There had been no time to pay on departure. The guy couldn't make change for a ruby. And Norm didn't want to advertise that he had several jewels on him, embedded in the tekryl waistband of his boxers. It had been tricky enough to get the stuff to melt over a gypsy camp stove in Memphis without burning up his shorts; now this was his first try at getting a stone back out. But Norm pretended he knew what he was doing: he borrowed the guy's packing knife, turned away discreetly, slit the waistband to loosen one, and gave a whole ruby to the driver. "Just pay me back some day," he said.

**

It was Alexis who picked Norm up in downtown London, at a coffee shop. He was wearing his Berkford disguise and she had a newer, stranger one, so at first they had a time trying to find each other among all the other people without attracting attention.

"Norm, I didn't want to scare you but they were closing in on you! At the campground, then at the harbor, they missed you by about one minute."

Alexis seemed to have decided to tell him everything possible before they got to her parents' house. Good idea, actually. He imagined that being around her parents would be pretty distracting. Possibly even horrifying, but he wasn't going to think about that. He was staying busy just trying to listen as fast as she was talking.

"Who was it?"

"Don't know who sent them: guys in a black car. The boat residents told about it afterward to the friend of the friend of that guy with the jet-powered delivery truck."

Once they were near her house they stopped on a dumpy-looking side street and hastily removed their disguises right there in the car. Norm sent Malo an unglamorous holo of him and Alexis, String in her carrier, and Ahanith's goatskin with a message, "Landed OK, we're all fine."

On second thought, he copied the holo to Ozzie and Freya. They'd find out sometime, anyway, that he was stopping in London at Alexis' house. No sense creating bad karma.

When they arrived at the house, Norm saw Alexis' uncle sitting out front, smoking. Uncle Yong waved cheerily. He looked like a guy who might have a sense of humor.

[I get out here,] String announced.

[We all do,] Norm answered. He opened the passenger-side door.

[I get out of this *box* here,] String insisted, with elaborate patience. When Norm finally complied, she stopped on her way across his lap to say: [This is the place where you have permission to be away from me.]

And she was gone.

**

Alexis told Norm her parents had been properly prepared to meet her robotics contest partner, and Mars greenhouse partner, Norman Garcia. That included being notified about his "eccentric sense of humor." All they had urgently wanted to know was whether he was being extradited, and with that settled he was a welcome guest, for a little while anyway.

Alexis would sleep on the living room couch and give her room to Norm, her parents had agreed with her. Norm refused, embarrassed, on the grounds that Alexis was a student who needed good rest and study space. Mrs. Wu accepted his generous offer reluctantly. And it seemed that saved embarrassment for everyone.

When Mr. Wu addressed a college question to Norm at dinner, he must have expected a standard answer. "Well, Norman," he said, "what do you feel is the most significant thing you have learned this semester?"

Norm considered and said: "The dormitory janitor taught me how he stays so happy."

Mrs. Wu laughed first. Norm could tell she had decided that this must be one of the droll things he would say. The rest of the Wu family laughed a little, anxiously. Except for Alexis. Norm watched her lopsided smile travel up one side of her face.

**

Norm had offered to do the dishes, to impress Mrs. Wu. That worked, and better yet, Alexis joined him to show where everything went.

She said softly so no one else could hear, "Norm, that cat you brought has met Tom."

"String? Oh. I forgot Tom was here."

"He's been hiding out. Feeds himself. No one knows this is his headquarters, except me." She giggled.

"Well, String is pretty persnickety. No one gets away with much when she's around. Has she told him off yet? Or vice versa?"

"Oh, no. Not at all. They like each other...They're out there conversing and plotting, thick as thieves."

Norm had an errand to do: needed a couple of things from a local store. It was their chance to escape and talk freely. And for him to get his arms around her. Some things had changed for him, he realized, and he had missed her. Seriously missed her. Scary.

In the batteries-and-electronics aisle Alexis went over the Oxbridge results, and showed Norm the actual message from Diana about how doubling live voices had resulted in a five-times reduction in tremors wherever it was done.

"You said that, and she told me, too. But sound is sound, right? I still don't really get how it makes a difference whether it's live or mechanical, or any of that stuff—"

"—Whatever works, works, right? For example, in the theory of relativity—"

"—Ugh, spare me. There has to be a way to amp the sound up so it just overpowers whatever is fracturing the crust—"

"—Yes, you would think so, but time is so short that all we can do is test things and use the theory that works best—"

"—I've been doing that, as you know: by getting the players out all over America. But jeez, there has to be a better way—"

They argued happily and ended by sitting at a coffee shop table drawing diagrams together.

"Well then," he said, and the rest was whispered because this wasn't something for others to hear, "why don't we do this, just the way I did in America, as a flanking action—like in a battle, you know, something to support the main action—and get players with slivers to campgrounds here in Britain. There have to be gypsy camps all over Britain, right?"

"Probably. Some are called Travelers here..." But Alexis didn't know for sure. "Who could we even ask?"

**

String told them: [One outside the city. Another near the Standing Stones—]

"Stonehenge?" Norm whispered.

Alexis nodded.

[—and other places. Names you don't know. It is possible to find them…]

They sat on the flagstones, shivering, in the courtyard outside Mrs. Wu's kitchen door. String stood before them with her tail tapping a tiny rhythm on the stone beneath her. The courtyard lights were dark, turned out to keep their presence a secret while Alexis' family was getting ready for bed and Uncle's family, on the next floor up, noisily did the same.

Alexis was saying, "If I weren't in school, we could make a tour of it, go to Stonehenge, then north to the lake country, and then…"

…Then this, then that; she seemed to just like chatting about it, and aside from being a little cold, he was happy to let her. "Ouch!" He looked down. String had dug all ten needle-like front claws into the toes of his hiking boots. *Thought these boots were impenetrable.*

"Shhhhhh…" Alexis warned.

He gaped at String.

The cat held Norm's eyes for a long moment, then eased the claws back out of his toes. She said, [*This is what we must do…*]

CHAPTER TWENTY-EIGHT

THE EARTH TREMBLED ALL NIGHT THERE in Seydisfjordur, in the harbor and beneath the hostel.

In the night, after Freya slept like a swimmer deep in dark water, she rose to the surface to find a clear sky and stars that shone on the busy Old Harbor in Reykjavik, which had moved somehow so it was right here. It was full of pleasure boats strung with colored lights. She lay on her back and floated and the water began to steam. It refreshed her so much that she dove deeply again, drawing strength from the pressure, the heat, and the rich minerals. When she resurfaced she dreamed again:

It was nearly dawn. The hills were jagged, outlined by the moon that had just dipped behind them. She could see trap-plants, whose lush flowers must be full of insects by now, closing for the day to digest. The air was heavy with their perfume. One by one the glowing lizards alighted on branches and their glows winked out.

As she and her Other walked along, singing from habit and no longer from inspiration, vehicles began to light the red dawn sky, rising in dozens from the port. Their mission, she suddenly knew, was war. Knowing filled her with terror.

A thundering impact threw them off their feet. As she fell she saw flamelike lights stabbing the sky and nearby earthen walls sliding into the water. The fall knocked her breath and the song from her.

Freya woke, wet with sweat and tears—still struggling, as she came up out of sleep, to get up again and sing. [Ahanith?] she called silently, without thinking.

Wherever Ahanith was supposed to be right now, she felt the Singer here with her, and the dream was one they had just dreamed together.

**

November what? Ozzie wasn't sure.

They had fished their way to Iceland, for about two weeks it seemed, while the days grew shorter and the nights longer. The clock displays on the ship weren't working. Their phones weren't either. He had lost track of time, unless it could be measured in fish caught, bait cut, fish stored.

But now Völun's hold was full, and he seemed to be

happy with the bargain he had made with them. He had never commented on their disguises, although they were sure he could tell the hair and beard were fake when he saw them close up. They had kept their part of the bargain, and Freya said she was sure he wouldn't betray them.

The guy could sell his cargo at one of the fish-houses here in Seydisfjordur, this port on the eastern coast of Iceland. The place was a few large handfuls of twinkling lights reflected in harbor water, ice and snow. Probably Völundur would shower and sleep at fishermen's lodgings, then go back out to sea again. He had pointed the way to the bus station and waved as they left him standing in the harbor lights by his fish-crammed boat. Behind him tall, bare-looking mountains rose, on either side of the harbor, into the darkness.

They had taken their fishy laundry to a laundromat, checked in at the one hostel in the town, and showered excessively to get the fish smells off; then they had both passed out discussing what to buy for supper.

The earth trembled all night.

It was dawn now, but 9 a.m. according to the clocks and their phones. They were gliding westward on the hoverbus from Seydisfjordur to Reykjavik, across a snowy and wild-looking landscape, with phone reception sketchy. Better than the reception out on the water, though.

So this is Iceland. He eagerly took in all he could see:

rugged plains, and mountains faintly visible farther away; all silver and black, very elegant in the gray light. It looked a little like New Mexico, the way it was out in the open country, but without as much growing here—except where it seemed to be greenish around steaming pools. An odd place: hot and cold at once, green and bare at once.

One by one they both pulled off the layers of sweaters they were wearing. The heat in the bus was much more than they were used to.

"Look," he said. Some more delayed messages trailed in, including the latest from Diana about PII findings.

Freya turned from her phone. "You getting this one? Yeah, I'm reading it. That private investigator was a spy? Creep scum! That has cost a month of investigation time at least!"

"No wonder Norm was being chased." Ozzie scowled. He and Freya had also just received the latest from London. *How will Norm get to us if they're following him, too, now?*

Diana said PII had captured the encrypted files that were sent by the two-timing investigator to Raker and GG. Still decoding them, she said.

But PII has been doing its part well: the technical part of the investigation. Based on my hunch, our team has checked and found that the technology being used by the "testing" company in OK is

creating vibrations exactly according to the ring formula. Not just like the formula. Using the formula *exactly*.

Freya nodded at him, with a look that said: "We thought so, didn't we?"

He sent a message back, hoping it would reach her faster than Diana's had reached them:

And they say this is oil testing? What are they doing it for?

The answer came quickly:

Exactly. It raises more questions because it makes no sense. But PII is a scientific investigator, not a people-chaser. We've hired the help of a new private investigator to look into the people and connections behind this. This one will be accompanied by one of our toughest staff 24 hours a day so he can't be bribed, or neglect the speed we need.

They sat in silence for a while. The rising sun cast rosy light that made the sweeping snow-and-rock landscape enormous, like a stage for heroes.

She said, "It's lucky Völundur was right: that there really is a bus." It had cost most of the money they had left, but fortunately no more.

He said quietly, thinking, "Because we got on an

Icelandic boat in Thurso, those guys who were chasing us probably have guessed that we'll end up in Iceland somewhere…It's a good thing people can't be traced by phones the way they once could."

"I don't remember: why were they tracing?"

"Because they could, I guess."

"Maybe it's lucky the towers were wrecked in the Wars."

"I think it was deliberate. Better anyway, having things connected through the worldweb. Connected pretty well, anyway." His phone had just cut out. Maybe it was the volcanoes.

CHAPTER TWENTY-NINE

THE VIBRATION WENT ON WITHOUT STOPPING, like in Oklahoma.

Ozzie set his pack down silently and looked around. They didn't dare turn lights on, but framed by little windows, snow swirled outside in the glow of the street lamps. A little of the light leaked in here, into the kitchen. The place looked as snug and tidy as the inside of a boat. But it didn't smell like a boat, thank the stars. It smelled like licorice.

There was a coffee mug on the counter by the sink. They had seen no vehicles outside, and no tracks in the snow leading to the door. Except that now there were his and Freya's tracks there, Ozzie thought. It would be too easy to spot that someone was inside now, even with the lights off.

Freya held her finger to her lips and moved out into the other rooms to check.

She returned. "No one here. But my mother's clothes

are here, and your father's."

"They may be back soon."

He went out to the dark walk at the front and flung a dish towel back and forth to scatter their footprints as he backed up the walkway and into the house again. He locked the door.

He turned on his phone light and set the phone on the counter so it shone softly against the kitchen wall. She put a kettle on the stove; they sat in the dark catching their breath. They drank some hot tea, a kind Ozzie had never tried, but it was pretty good anyway.

They got their phones to download completely at last.

"Newsfeed," Freya said without emotion. "Look: Raker now wants Iceland to extradite us! We have been traced here, it says, and he's pushing at the Icelandic government to turn us over."

"Good thing our parents aren't home. Got to keep them from knowing exactly where we are," Ozzie said.

She nodded. She opened the refrigerator, sighed, and shut it. "We'd better not eat anything that they might notice. But here—" She pulled two tekryl tins of fish from a pantry shelf.

He nodded. He was getting used to fish. And he was starving.

They washed the tea mugs carefully, dried them, emptied the kettle and returned everything to its place.

Then they left to walk a mile to the Reykjavik Hostel.

**

It was still dark when they both found themselves awake next morning. "Such a relief to sleep," he said. She agreed. She didn't mention that shaking had wakened her again and again all night long. "But we don't have the money to stay here much longer, right?" she said.

"No." He bit his lip.

"Text in from Diana," he said. He laughed, for the first time in a week at least. "She says she has sent documents to the British and Icelandic governments showing that we have been employed by PII as researchers... and travel to Britain, then Iceland, has been part of our jobs..." He laughed again, and read: "Please send address when you can for paychecks from recent weeks to be forwarded to you."

"Perfect timing. We need the money!...But we have no address," Freya said.

"Yeah. Funny, huh?"

**

Ozzie went out for a walk when the sun began to rise. He could tell she wanted more sleep, so he told Freya that he needed to get out and look around. Which he did.

He went in disguise of course, wearing the curly dark beard and knit cap pulled down almost to his eyebrows. *Even Dad and Ilse wouldn't recognize me.* Good to be out again in the cold, clean air.

The gulls were crying and making circles above the

harbor as he walked by the boats. He could feel again the joy of flight, just by watching them.

He loved flying. In his not-going-to-happen internship with Grand Galactic, the training would have included a lot more of it as a ramp-up to piloting the full-sized trading ship: first some small exotic craft, then a few big training airships, before takeoff in *The Liberty*. Then, after takeoff, he would take time at the controls. Gradually he'd be trusted to fly *The Liberty* herself under the eye of the captain.

He tried not to think about his trading ship. Which by now must be outfitted and almost ready for her grand nighttime procession from the factory in Las Cruces northward up the highway. He saw her, enormous in the darkness, tugged by heavy trucks and escorted by flashing lights and guard vehicles all the way to her gate and berth at the Spaceport.

He shook the image away.

He looked instead at the fishing boats tied up here, rocking in the water, thumping against the dock.

One had a sign on it. He couldn't read that, or the name of the boat, but the words were followed by a number and the symbol for Icelandic money, krona. For sale. *Wonder how much that is in dollars.* It was about the size of Völun's boat, but older and it would need a lot of work and new paint to look decent. He took down the address and number on his phone, for no particular reason. Just kind of nice to think about owning a boat.

Under his scarf, something was burning the skin at his throat. He felt for it, absently. The ring was very hot. Because of the scarf, maybe. He loosened the scarf and walked on in the sunlight.

Then he stopped, sat on a catwalk with his legs dangling above the water—low to stay out of view—and called.

**

It was then that Norm called in.

"Norm! You OK? Where are you?"

"Reykjavik, buddy." Ozzie could hear his wise-guy grin.

"What!?"

"I have a disguise on, so I won't blow your cover. Tell me where you are and I'll find you."

It had been a strange couple of months. Ozzie told Norm where he was, then hid nearby so he could make sure Norm came alone.

His fingers went to the ring, still on its chain around his neck. Hot again. *Why is it hot? Must be from my body heat, soaking it up... But the thing wasn't hot when Freya wore it. Was it?* He'd ask her if it was.

After Norm stood waiting for a few minutes, eating something from his pack, Ozzie poked him in the back and grinned at Norm's surprise.

"Gotcha," he said.

Ozzie asked, so Norm told about his detour to

London. "I knew I oughta get here fast, but Alexis and I tanked up on togetherness while I was there," Norm explained to Ozzie. "We needed it."

Ozzie rolled his eyes and grinned, glad to see him. "How'd you get here?"

"Got a ride with one of the Oxbridge boys who was heading up to Loch Something for a golf tournament. I had to learn to play golf to get the ride, but the food and beer were good. I turned out to be a really bad golfer—what a corny game that is—but he didn't seem to mind."

"And... how'd you get from there to here?"

"The same guy had another friend with a little private plane, there for the tournament. He didn't want to take me along, at first, but really he needed someone else to fly with him because he was a little drunk and someone had to keep him awake. He taught me to fly, Ozzie!"

Ozzie gaped. "And you two landed OK."

"Yep. A little hair-raising but...tell you about it sometime."

**

Norm was willing. So Ozzie called the boat-owner and started the conversation again at the place where they had been when Norm rang in.

"Motor good... Doesn't leak... Four little berths? And a heater and lighting inside..." he nodded to Norm.

"...Why are you selling? I see...And what's the price in

U.S. dollars?"

Ozzie closed his phone. "He's coming, about ten minutes away."

"The boat's a clever idea, buddy. Especially since Iceland could agree with Raker to extradite you—like any minute. But how you gonna pay for it?" Norm said.

Ozzie held up the ring on the chain.

Norm whistled.

**

After greeting Freya warmly from somewhere near the goatskin, Ahanith had drifted away. But not to sleep; she seemed to actually be gone. Maybe she was out looking at the harbor.

They stood in the hostel common room while Ozzie and Norm took off their coats and unwound their scarves. Norm's pack lay on the floor at his feet. "You're not mad at me, Freya?" he said. They found places to sit, facing a wood stove.

"Call me 'Francie' here," she whispered. "No, I knew you'd go see Alexis first." He seemed crestfallen at being predictable as well as unreliable. She smiled at him. "It was a good idea, Norm, actually."

Ozzie knew how Norm had enraged her on Mars, but by the end of that quest his dedication and his idiot humor seemed to have won her loyalty forever.

"So how did you like college?" she said.

"It was a little odd," Norm replied.

They waited.

"Until I decided that I would just learn something anyway."

Ozzie waited some more. Punchline coming. Freya was already chuckling softly. Cocooned in a thick wool sweater, she looked like she had finally slept well.

Norm sighed. "My professors didn't seem to have learned anything very recently, and they were handing out info I already knew. Some of it I already knew wasn't even *true* anymore. Old outdated data. So I was trying to figure out what to do there. Then I noticed that the people I was really learning from, like my roommate and the janitor and the research librarian—"

"—the janitor?"

"—yeah, he was the happiest fella I've ever seen, guys, no kidding. Listen to this: one day I asked him why he was so happy. He said, 'I like being a janitor. I like cleaning things.' Cleaning things. Wild, huh? The guy has no family. Lost his son in one of the Wars. And he could be One Sad Pup. But he was always happy."

Ozzie blinked, thinking about this. Especially, thinking about Norm thinking about this.

"I asked him again one time how he did it, Oz. Arthur. And he said he always enjoyed working, always tried to help people out when he could..." Norm counted these things off on his fingers. "Said he never decided he had to own certain things to be happy, and he liked to read. It was easy to be happy, between his work and always having something from the library to read."

The opportunity was tempting, but Ozzie didn't ask if this experience had made Norm decide to read or clean more often.

**

Freya fed them all in the hostel kitchen. Everything trembled while they ate.

Norm said, "You're nicer than you used to be, Freya. Francie."

"Because I made you lunch?" She raised her eyebrows at him.

"No, although this is better than most of the stuff you've cooked…"

"None for *you* next time! …Was I not-nice before?"

"Not this nice. Sometimes you were scary. No offense. But after all that volcano-shaking in Oklahoma, I understood what you were going through."

He grinned at her as a hasty afterthought. She couldn't help smiling at this, his latest revelation.

"String didn't come with you?" Freya decided to change the subject. She helped them all to more food from cans.

"She stayed in England, with old Tom. Hey, I have something to tell you; something Alexis and I did—"

"OK, but tell it later, Norm," Ozzie said. "We have something to show Francie, right?"

**

The sun was setting already. Stars showed through the sunset. "It's because we're nearing the winter solstice, Norm," Freya answered him.

"This is one crazy place," he said.

They walked Freya to the boat. Her eyes became large.

Suddenly it looked shabby to Ozzie; old tub of rust and peeling paint. It would be hard to sell, when the time came, because it would cost almost as much to fix up as a lighter, more current boat might cost today. With times this tough, no wonder the guy had taken one heavy gold ring as the down payment for it. "It's not beautiful, but Norm and I have tried it out and it runs well. The owner gave us his word on it, too."

"It's yours?" Freya was astonished, and more so when she understood the terms of the trade: the ring. Plus one tenth of their fishing earnings, every month, till it was paid in full. "But your grandfather's ring—"

"We've got the formula safely recorded. The rest is the past," he said. "I'm trading the past for the future. And look: it's a place to live without paying hostel fees, a way offshore so we can't be caught and extradited, it's a way to stay unknown to our parents so they aren't in trouble. While we find out what to do about the volcanoes."

"And, it's one more thing—" Norm began.

"Wait: do you know what this boat is called?" Freya chuckled.

"Signed the papers. The name is right there," Ozzie pointed to the bow.

"*Hringur*," she laughed at him outright. "It means 'ring.' You traded one hringur for another one." She hugged him. "Thanks for believing in me," she whispered in his ear.

He took her hand and walked her onto the boat. "Let me show you around," he said.

Norm said he'd go look for some food somewhere.

CHAPTER THIRTY

THEY OCCUPIED THE BOAT RIGHT AWAY, moving their things into it from the Reykjavik Hostel. Freya asked Ahanith where she would like to live, onboard.

[Such a wet place...] She decided on one of the kitchen cupboards, clearly hoping it would be warm and dry there. Since Norm was already unloading his bag of groceries into a cupboard, Freya quickly put the goatskin on Ahanith's chosen shelf. It seemed like an undignified place for the Singer to reside, but she was content, she said.

"We bought the boat in Norm's name for safety," Ozzie said. "Not safe to use my legal name here. But Norm can't be arrested—"

"Norm can be harassed, chased, and threatened," Norm recited to Freya, "but not arrested—"

"—so he's the new owner and captain of this excellent vessel."

**

It wasn't until the next day that they caught up on the rest of the details. Ozzie woke in the dim, cold guts of the boat and went above to find Freya wandering around on deck in the gray light of almost-dawn with her tea steaming, hefting this rope and trying out that piece of equipment, admiring it as if it all were made of pure gold. He watched her for a long time before he even wanted to say anything.

But by then Norm had arrived on deck, too.

"Now I can tell you two my idea," Norm said. "I think we should set this tub up to broadcast underground radio from the North Sea. It's an honorable tradition, you know."

Ozzie just looked at him. This happened to him now and then with Norm: he had no idea what the guy was talking about.

Norm explained about the unlicensed radio station a hundred years ago, broadcasting some kind of unacceptable music from a little North Sea ship. It was a historic act of defiance that had a certain appeal. "We should broadcast the Singer, live, and her music. And whatever else we want," he said.

"From this little boat?"

"Sure. The equipment is small. Radio is old technology, but it's cheap and it works, even where the worldweb doesn't. I could set up, umm, on the little galley table downstairs, with about a cubic foot of electronics, and we'd be broadcasting our brains out."

"Maybe not the table, Norm. We have to eat somewhere. But broadcast the Singer? Brilliant! What do you think, Ozzie?"

He agreed. It was in fact brilliant.

**

First priority: a training session on what to do with this boat. Ozzie and Norm showed Freya all they'd learned from the former owner. Fishing Boat 101.

For the 201 course, Ozzie and Freya showed Norm the other piloting and navigating they had learned from Völun. Then each one took a turn moving the boat out of the harbor, out to the open sea, and back, a little further each time.

Norm said, "At least here on the water everything isn't shaking all the time. Well it's shaking differently anyway…Hey, what do we do when we run out of fuel? My cash is down to a few bucks."

"Ours too. We'll catch fish, with nets and lines, and sell them," Ozzie had already planned it.

Freya said, "I'll show you how the fishing stuff works. We've had practice with Völun." She wrinkled her nose at the experience, and chuckled. "We can do it, Norm, don't worry. If the worst happens, we can eat what we catch and buy fuel. Better, we can sell some for things we need, like tools and vegetables."

"For now we mostly have to look like we know what we're doing," Ozzie said. "In case we get stopped. We'll get better at it."

"Cold business to be in. Maybe some people are going to think it's weird, that three people our age are running a fishing biz?"

Freya shrugged. "In Iceland these days, the people who have stayed do what they must."

When they toured the kitchen and showed Norm how the stove worked, he hefted a frying pan soberly, then astonished them by announcing that he'd try some cooking while they sailed: "I learned from my roommate to make a couple of good things. But no cooking while I broadcast live, alright? And we have to get some of the right kind of spices."

Norm was delighted when Ozzie and Freya agreed to get the radio station electronics next. Freya promised to show them around Reykjavik a little as they went.

Norm always seemed to make electronic parts appear out of the air, so it wasn't a big surprise to Ozzie when their first stop was a jewelry store to cash in a little ruby from Malo. "Gave it to me to help the Singer get her sound out," Norm said.

Malo's hand in this again; did he see into the future?

In addition to the electronics, they also bought heavy jackets, second-hand ones, for each of them. And a spare. "Good idea, Norm." Freya put hers on right away.

In spite of some spare jeans that they also picked up,

when they returned to the dock Norm had enough left from the ruby for fuel, so they filled the tank at the port store too and bought a bag of groceries: bread and butter, dried beef, cans of things.

We are fisherfolk now. This is all pretty crazy, Ozzie thought. *But here goes.* "Freya, since Norm is the captain, how 'bout if we let him choose our first course?"

**

Norm was sure the course he had picked was a smart one: the *Hringur* ran east and west and east again, off the southern Icelandic coast from Rekjavik to Seydisfjordur and back, with Ozzie and Freya and the Mystery DJ disguised as fishermen and fish-preparers. The DJ played "vintage and alternative sounds" from the waves to Britain, Scandinavia and the U.S., giving all-night broadcast time to Ahanith's song. Once a day the Singer sang live, amped up as far as Norm could make it go.

Within three weeks, Norm's irresistible voice—*well, it seems to be,* he thought—and his program, MuckRaker Radio, had become familiar to British and European listeners, emanating from mysterious, ever-changing locations on the North Sea. He had borrowed "MuckRaker" from his usual old sources just because he liked the sound of it.

He gave reviews of vintage flat-films and repeated such slogans as Save Iceland, Vanquish the Volcanoes, and Sing the Song, Stop the Quakes.

The Singer's song began to be a cult favorite, at least in Seydisfjordur and Reykjavik. *It might even go viral*, he thought.

As they sailed Norm set up little chunks of robotics to run the broadcasting station 24 hours on an hour's worth of planning time, so the three could haul the fishnets and empty them, prepare the fish for sale at each port, and he could get some sleep too.

And drink tea to warm up. Even Norm drank the watery stuff, but he wasn't sure he'd ever be warm again. "Warm is overrated, Norm," Ozzie said. Easy for him to say; he had more practice with cold.

On the air, Norm urged American and British radio stations to play the Singer's melody, too, but that didn't seem to be working because he didn't hear it happening on any of the other stations. But who could tell, with the trashy reception they had out here? Maybe that was because of the volcanoes.

**

Norm was getting good at piloting the boat, and Freya already knew how, so Ozzie wasn't needed as much of the time in the wheelhouse. Cutting bait was boring. Unfortunately these factors meant that Ozzie had time to think.

Today it was on the news, on all the radio they could receive, that they were under threat of extradition from Iceland now, as well as England.

Today a cold wind blew them roughly around on the sea as if they were trivial and drowning them would be no big deal. The wind turned the peaks of the waves to spray. Even when he wasn't steering the boat, Ozzie was busy worrying about whether the boat was being steered safely enough. Maybe *no one* could steer it safely enough, and the sea would just swallow them.

Right now he sat warming at the little table beside the galley. He looked at Freya. Everything stank of fish in here, even to his numb nose. He said, "Is it my imagination, or is everything just getting worse?"

Freya surprised him. She slid over beside him on the seat and gazed at him eye to eye until he looked at her directly. Then she said, "No, it's getting better," and she smiled merrily at him.

He laughed. How did she know to do that?

She took his hands and kissed the backs of them. They must smell like fish but they weren't as cold as hers. He held her hands in his to warm them. How did she know that was just what he needed, when he didn't know it himself?

For faster warming he unzipped his jacket neatly and pulled her cold hands into it at his sides, then pushed his own hands into her coat and around her back, to draw her closer so they shared the two jackets together. The warmth, after so much cold wind, made him drowsy. A few minutes later he woke and said some blurry words into her hair that meant "What if we stay this way till it's my turn at the wheel again?"

She nodded and sighed in her sleep, her forehead leaning against his shoulder.

**

Freya asked Ahanith to make more of a presence so they could see her; it seemed to injure her dignity to be walked through or ignored when she was actually right here. Ozzie listened in.

Ahanith answered: [Takes a lot of energy to create such illusions. Energy is still returning to me.]

[Maybe just a simple version, something we can see?]

She showed her green eyes.

Freya nodded judiciously. [That will work.]

**

They had returned westward and anchored in Reykjavik harbor to sell their fish and buy more supplies. Here in port they covered the name of the boat, casually, with draped nets or rope or laundry and wore disguises everywhere they went in town. Raker's bad news was on all the stations. But the Singer was on all of the stations here, too, Norm said.

The sun's reflection in the water had faded an hour ago, and now it was dark and cold enough to be called night. Ozzie and Freya sat on the big equipment chest at the back of the wheelhouse, their feet on a coil of rope.

The afternoon deck chores were finished but they

were both reluctant to go below. Below Norm entertained his radio audience as he queued up the next music. Here, it was quiet. And at this end of the harbor it was fairly dark. They caught glimpses of northern lights moving across the blackness, like curtains of rain rippling across the desert back home.

Ozzie was describing the captain's quarters and the bridge in a trading starship: the tidy cabinets, the compact shelves of books and tools—so similar to a ship that sailed on water, like the boat they were on. It seemed to make her happy to hear about it: a place for everything, everything stowed neatly.

The Singer's eyes surprised them, appearing above the rope coil. She had left cupboard and goatskin to join them.

[The evening is beautiful.]

They agreed. Ozzie was still a little awed by her. They waited in silence.

[I could tell you about a night like this, only dryer, warmer. Martian…when I was joined with my Other.]

They listened. As she told the story he remembered parts of it from dreams brought by the cats. Now it was like a dream that more than one was dreaming.

When Norm came up the ladder they were standing on the ice-cold deck, so entranced by her tale that they weren't even very embarrassed to be found holding ankhs and flowers—and maybe they were imaginary, but Ozzie was sure he could smell the sweetness of the flowers—as she sang to them:

> [We are two
> We are two who walk as one
> Walk as one for all our days
> Together
> All our days
> We walk as one
> Journey side by side
> Together
> We are two who sing as one
> Sing as one for all our days...]

She stopped. [Is Alexis coming here?] she asked.

Strange question. Ozzie answered: [No, Ahanith, she...]

But Freya's hand stilled him. She looked around them and pointed at the lights of a small motorboat cruising slowly in the harbor nearby. It looked as if it were searching for something.

"What's up, Freya?" Norm said. "If you've decided you don't want to marry him, I don't blame you—"

"Norm, get up on something and wave your jacket!" she said. "Or whistle."

CHAPTER THIRTY-ONE

THE HORN OF AN INCOMING VESSEL sounded deeply far out there in the dark, in the cold, salty-smelling air. The little lighted motorboat was streaking away, leaving them here to witness the impossible: Alexis stood in front of them with her backpack on, her fishtail braid wound into some kind of big knot on the back of her head.

"The Snowdon volcano in Wales is smoking from a crevice, did you hear? That one is so old it could have mold on the smoke! That was the end, guys. I told my parents that I needed to leave school, and London, to help…

"When I told them Oxbridge would have to wait, they were *furious*. So I said, 'What good would it do for me to get you a new house if all of Britain becomes a volcano?' …They didn't believe me." One tear slid out of the corner of one eye. Just one.

Ozzie had never seen her look so alive.

"I may be wrong to do this. But you know what? It's my decision. If they decide for me, I will always be a child."

"Go Alexis!" Norm grinned. Freya took a turn next and hugged her.

"…And now that I have decided, I will make my decision succeed, of course. So it will turn out to be the right decision after all!" she smiled brightly.

"My dear, your reasoning is staggering," Norm said. "…However," he added in a loud whisper, "now you must excuse us; we seem to be in the midst of a mock wedding."

In his best preacher voice, he restarted the ceremony: "Dearly beloved from both planets…"

But when Ahanith began once more, from the beginning, to sing her wedding song, it seemed to be hard even for Norm to successfully resist her spell.

Again Ozzie's hands held an ankh and a flower. Tears ran in sheets down Alexis' cheeks. Norm wagged his head at her. "It's *pretend*," he whispered. "Don't cry."

"But I love it. Happy tears, Norm."

"Happy tears." He took a fingerful off her cheek and tasted it thoughtfully. He shrugged thoughtfully too.

In spite of himself, by the end of the song he was so entranced that he almost missed his cue. He barely recovered in time: "Ozzie and Freya, I now pronounce you unmarried Earth-people. You may kiss each other. As if you haven't done that plenty already."

**

Norm and Alexis sat at the little galley table, taking their turn at breakfast, while Ozzie and Freya quickly pulled their topside clothes—boots and jackets, knit hats and scarves—back on. Alexis and Norm had done their turns on deck together until both were good pilots, and all four had enough confidence now to leave the boat on autopilot for short stretches.

Norm said it again another way: "Alexis, think with me. Isn't there a way we can amp up the sound on my station so it's more effective? Like by adding an aftervibe to the vibrations or something. I still think there must be a way to do it."

"Maybe we could test some ideas with the local cats," Ozzie said, digging into his pockets for his gloves. "Cats seem to hear Ahanith well. Too bad String and Tom didn't come along... Why did they stay in London?"

"They didn't. Didn't I tell you? Alexis and I set up those two to do the grand tour of England and take some players and slivers around the place."

Alexis said, "We took them by high-speed to the London campground—a pretty wild place, my friends—and searched out some gypsies and their Traveller friends there, gave them a bunch of slivers and players to distribute. We designed a harness for each cat so they could each carry a player and two master data slivers with them to Scotland or Ireland if they wanted."

Freya froze with one boot half on and gazed, imagining the two cats off together on this trek.

"—Not heavy for them, Freya." Norm countered any arguments in advance. "About five ounces total. Good to get them busy, you know, so they wouldn't make Alexis' parents mad by hanging out around the back garden too much.—And so they wouldn't follow me, either." He grinned at Ozzie.

**

The sky was blue, with fat clouds, and the waves had froth on their tops from the wind. Ozzie heard a moan of despair. Holding the wheel, he leaned head and shoulders out of the wheelhouse to look. There Alexis stood, cutting fish at the table back there on the rocking deck of the boat. Norm was emptying the last of a bucketful of fish onto the surface.

"Look at me," she lamented. She wiped fish entrails from her knife across the front of her apron. "I've become a fishwife!"

"Hey, that mean you're married to a fish??" Norm grinned. She waved the fishy blade at him threateningly. The boat lurched and she lurched too, right toward him like a fencer.

She clutched the front of her apron and paled apologetically.

**

Sometime in early December, Diana notified them:

I don't know exactly where you are and I'm not asking. But I have been listening to Norm's broadcasts. In fact, PII engineers stay tuned in to check on seismic effects caused by the song.

We're still testing to see what affects the Oklahoma volcano. To answer your question: addition of your broadcast boat makes no difference in Oklahoma, sorry to say. It does make a small difference at our Icelandic test station. But the Oxbridge Chorus gave a 1% improvement in both locations.

Try getting more people to sing it live!

I'll be in Iceland for some testing soon.

**

Norm was getting used to the smell of fish and salt water and the roll of the boat. He could walk to the bathroom, cleverly called "the head," out of a deep sleep without losing his balance.

He had just set up today's playlist, all loaded for the next 24 hours. He stuck the sliver into a slot in the little cube of electronic broadcasting gear. It sat on a counter next to the galley table, where most people might put a toaster or something unimportant like that.

Then he switched the station robotics over to automatic play so he could do his next watch and fishing duty. The robotics would mix a bunch of intro statements and recordings of Ahanith's song with artsy selections, flat-film and holovid reviews, and vintage rock cuts so it could all run for hours without him, and people could sing along with the Growing Song whenever they were listening, anywhere. He hoped they were listening somewhere.

He was having a ball broadcasting, even if his ratings from PII were still not good. And it didn't seem like his listeners were singing much. Or loud enough. He wasn't even sure how many listeners he had, actually.

**

Alexis understood the problem. While they cut fish for bait together at the back of the boat she listened and said, "I think Brits and Americans have a little trouble with this music. These are not familiar sounds to them." She guessed at who the sounds would be familiar to: Native Americans, Arabs, Middle Eastern people, Asians, Hebrew-speaking people. "There are similar sounds in their music I think."

"That's it. I need to get it to radio stations in other places."

Something was coming to him.

His hands were freezing and fishy from cutting bait.

His pockets were slimy, icy and fishy from putting his wet hands in them to warm. It was Alexis' turn at the wheel, then his. Ozzie and Freya were hauling in the lines. Now, while he watched the prow go up and down on the gray waves, Norm thought his idea through:

There were the remaining rubies from Malo, which he had sealed into the stretchy tekryl waistband of a second pair of his boxer shorts. He hoped he still had the shorts, after all these weeks of doing laundry in the port coin washers.

With those jewels he could fund his idea. It would be a contest, for any radio station anywhere, with cash prizes supplied by him. Well, supplied by Malo and his rubies. He himself would just have the fun of giving them out. But who didn't love cash prizes?

The stations would get more interest from their fans because of the prizes. The contest would involve listeners singing the song and sending in sound-slivers for their local stations to judge. That meant lots more people in lots of places—Vietnam, Estonia, Brazil and Mozambique—singing Ahanith's song, because people would need to practice first, he figured. The entries would be played on the local stations—more airtime for the song. And that would inspire more singers to try.

Single singers or duets, choral groups or tribal groups could enter; it didn't matter. Each station would pick its very best entries as local winners to submit to him, the Mystery DJ. The local winners, contest action and results would be announced regularly on his station, so that would get him more listeners too. And he could give lots of broadcast time to playing the entries, all the different versions of the song.

Norm would judge the final winners based on Diana's mercilessly accurate volcanic measuring devices: whoever calmed down the shaking of the earth most, won. Period. The stations whose top choices won the first, second and third prizes would also get prize money for the station. But the winning singers themselves should get the most prize money, he decided.

**

"Money in the till, makes good will," Norm grinned at Alexis, feeling rich and generous when he explained his scheme to her later. "Now this has to be some powerful good karma, getting this contest out."

"What's all that about karma?" she said. "You keep saying that."

"Part of my college education."

She gave him a lopsided smile and returned to what she was reading. She was reviewing and editing his announcement to the station owners, contest instructions for the DJs, a ready-made contest blast for their listeners. "This one sounds too much like hype, Norm. How about just saying—"

"Sure. That's fine. Best part is, these station owners can get it all electronically. I don't have to haul the players and slivers myself or make backpacks for cats. Just ship a bunch of electronic digits to—how many?—68,000? radio stations in the world? Do you for sure have the whole list of the stations loaded?"

"Actually 65,581. I do. They are all loaded on your phone, ready to go as soon as we have these messages and the music loaded..."

It was already time for them to do deck-duty again. But first Norm went into what he called the "bunkhouse" to make sure he still had the right pair of boxer shorts. Then he remembered something that he wanted to ask Freya, and he was on his way up the ladder to locate her when a text came in.

CHAPTER THIRTY-TWO

EARLY DUSK HAD COME. Norm stood on the deck alone, looking out to sea and being quiet.

Clearly something was wrong. "Norm?" Ozzie said from the wheelhouse. Alexis and Freya clattered up the ladder from below.

"Hey," Norm stuck his head in at the wheelhouse door. Ozzie cut the engine and put the boat on autopilot.

Ozzie had to pry but Norm finally spilled it: "Just got an e-transmit of a notice from Berkford: slimy Raker has goaded the California legislature to 'censure Norman Garcia' for 'assisting clandestine activities by enemies of the U.S. Government.' Ha! Ozzie, my friend, he accuses us of what he is probably doing himself," Norm finished in full-on preacher voice.

"But what—"

"The legislature was a good junkyard dog. They barked and Berkford threatened to remove my scholarship if I don't return immediately."

Ozzie swore.

"Norm, he's just trying to shut you up." Freya shook her head bitterly.

"Think your parents have heard about this?" Alexis said.

"No, they won't be paying attention to the news. They think I'm still at school."

"You never even told them you were leaving?"

"No."

"You get on the phone and talk to them, as soon as you can! They may know more than you think. How worried they must be! At least my parents know what idiocy I've committed to."

**

"I have finally got my newsfeeds in, Ozzie," Freya said. "Now that we're nearer to port."

"And?"

"Well, besides more about Raker's extradition plans..." She refused to read it. She handed him her phone and took the wheel from him.

A photo and the story: Seth Raker, winner of the GG Captain's Apprenticeship.

He just handed it back and took the wheel again.

Later, when he sat at the galley table mending a rope, Freya said, "I don't know, but I think we're going to win, Ozzie. In spite of the way it seems, you know, with Seth."

He shrugged.

Ahanith crooned in the cupboard. It was something she had begun to do now and then.

"I remembered just now: it was the Egyptian guide who said it, that even your enemies help you. How could that be? But maybe Seth is helping us to succeed in some way."

"Hard to imagine," Ozzie said. He went on repairing the rope, looking now and then at the instructional holo hovering in the air in front of him.

Since she was still standing there, he tried harder to imagine it. But still he couldn't. "That's OK, Freya. He'll get his."

The Singer crooned on.

**

Norm came to stand watch with him in the wheelhouse. "Damn cold out here," he said.

"Yeah. So what are you going to do about Berkford?" Ozzie said.

"Nothing. What are you going to do about Sethie and the apprenticeship? Freya told me."

"Nothing."

They sailed onward, east toward the fish markets at Seydisfjordur, in silence for some time. "Ozzie! Someone behind us, closing fast."

He turned and looked out the wheelhouse door. A ship, not just a fishing boat, but with few lights, was coming right at them from behind. Alarmingly close, and not slowing. "Get the lights off, Norm!"

Norm stared, but he hit the buttons and ran to the hatch to call down belowdecks: "All lights off, ladies! Now!" Ozzie's swerve to port was so abrupt that he heard Norm land with a thump on the tilting deck, then yell when he slid through the hatch and down the ladder into the galley.

Ozzie could feel the swell beneath them as the ship roared by, pushing them sideward so fast that the wave could have rolled right over their boat. But his swerve put most of it behind them just in time. The deck under his feet rose, went to zero gravity as it emerged from the drag of the water, and skimmed like a surfboard down the curl of an enormous wave; then the boat went to 3 Gs as it hit the trough of the wave and its keel sliced deep into the water again.

The *Hringur* bobbed and righted itself. By the time it did, the ship had already disappeared into the darkness.

**

"...Norm, I don't think you really even know what 'karma' means," Freya said. "Do you?"

Norm snorted.

Gray light illuminated the hatch from the sky above it. Ozzie had made breakfast. They were a number of miles out from Seydisfjordur, on the coast of eastern Iceland, with Norm here below to collect plates of food for Alexis and him to eat in the wheelhouse.

"Nearly time," Ozzie called. "Get disguises on, nearing port." It would be some miles yet, but there could be daytime boat traffic when they were this close to the place. Norm set the plates down again, hunted around and gathered up his disguise.

"Oh no," Freya said, and sighed. "Newsfeed. Another volcano report."

No one spoke, so she went on:

"Gujarat on the west cost of India, oozing and smoking...

"Tremors in Ethiopia—at O'a, an ancient one! And in Thailand! Quaking shook a temple down...

"Jorcada, ancient volcano in the mountains of Bolivia...cracking there at the base...

Ozzie groaned. He looked over her shoulder and pointed at the map on the holosheet. "But we also have everything quiet right now in Italy, Australia—"

"—and don't forget Antarctica, nothing much happening there—" Norm said. He grinned.

Freya made a disgusted face at him. "Wish they were ALL in Antarctica. Then our job would be easier, not so scary."

She thought she heard the Singer crooning softly from the kitchen.

"Emergency!" Alexis' voice came down the hatchway from above. "We're being chased!"

"Get your disguise on!" Ozzie yelled up to her, heading for the ladder. He pulled the soggy beard and hat from his jacket pocket and shoved them on any which way, then disappeared above. Freya and Norm followed his example and clambered up on deck.

Freya saw a big black-and-white boat coming at them from behind, closing in. "Get the name and ID numbers, Norm!" Ozzie called.

"But Norm, hold on to something!" Freya watched in horror as its huge prow grew taller behind them. She grabbed a metal railing on the outside of the wheelhouse. Through the window she saw that Ozzie had taken the wheel, and Alexis was hanging on to the wall somehow with her redhaired wig askew.

The ship, closing the distance behind them, must be ten times their bulk and weight. Freya knew it was about to go around them, one side or the other. Something about the motion told her: "Go left, Ozzie!"

And whether he heard her or not he spun the wheel to the left just as the enormous hulk veered to their right side, roaring by. Their boat arced to port; she felt a dizzy elevator feeling as they rose into the air on the crest of the wave with no weight and no water ahead of them, perched on the edge of nothing five stories up. Then they slid like skiers down a huge curling slope.

If we lose someone in this wave, we could never..."Hold ON, everyone!" she cried desperately, grabbing the handle of the storage chest with her other hand.

They hit the bottom of the trough, and weightlessness turned to sickening heaviness. Water poured over them all. *The hatch is open...*

They seemed to be plunging to the bottom of the sea. And then something pushed them just as hard, back upward.

"Hold ON!" She screamed again through the roar of water. The boat rose up out of the waves, bounced back in, rose and bounced again, and settled, bobbing wildly in the sloshing sliced-up sea.

**

"Yeah, I shut the hatch," Norm said. "You surprised? You've nagged me about it enough, buddy."

Ozzie thumped him on the back. He looked around them with the enoculars: No sign of any ship in the full 360 degrees. *That thing must have been powerful to disappear so fast.* "OK for now," he breathed. "My turn at the wheel anyway. Someone bring me food?"

Alexis hugged everyone. "Anyone get the name and numbers? No?"

"Maybe. I shot a picture of them as they went by," Norm told them. He clattered downstairs to look at what his camera had captured.

Breakfast was all over the galley walls. Ahanith was silent. But Freya was hungry enough to clean up the mess and start frying the eggs that were left.

**

Ozzie's legs came down the ladder.

"Ozzie," Norm said, "got a great idea that Alexis and I have been working on. Want to tell you about it—"

"—Safe in port now." Ozzie's head followed his feet down the ladder into the room. He put his empty breakfast plate into the sink. "Your watch, Freya, and we have to get the fish ready to sell fast."

"Oh boy, cash time. Groceries," Norm said. "We've got part of the name and number for that piece of junk that almost hit us, Ozzie. Just not quite all… Maybe enough to find out the rest, Alexis?"

She didn't answer. Alexis sat at the table with her head resting on her arms.

Freya said, "We also need to get some sleep while we're here in port. Whose turn to sleep, Captain?"

"No clue," Norm said. He squatted and peered at Alexis from the side, diagnostically. "Asleep," he said. "Must be her turn."

**

Well, this is more fun than school, Norm thought. Even emptying the fish out of the hold freezer was.

He added to his store of good karma by doing Alexis' share of the harbor chores, only waking her up for the fun parts at the end.

While they loaded their freshly-caught fish into the purchaser's refrigerator boxes, he remembered that he wanted to ask Freya something about selling the rest of the rubies. Then they were busy again, off to get groceries and pay for showers in the little fishermen's service station, and he forgot.

**

The load of fish was sold. The hold was empty. The cupboards and the small refrigerator were full.

With their newly-washed jackets rolled behind their heads as pillows, Ozzie and Freya, Norm and Alexis sprawled on the seats of the eating nook, their legs interlacing beneath the little galley table. Rest began to settle into Freya's whole body.

Ahanith crooned quietly, coming closer. [Can you hear me, each of you?] she said now. Her eyes appeared near the center of the galley.

Freya saw the others nod. Ahanith said:

[Now, when no cats are here, I should tell you this tale of two planets. While we have time, because another time you may need to know.] And without waiting for assent, she began:

[Long ago, many thousands of years more than you can even imagine, the planets you call Mars and Earth were occupied by very similar peoples. They were not at all the same, but similar enough.

[They traveled back and forth between the two planets, and took professions, worked, married, produced children, created beautiful things and were happy, suffered and died—much the same in both places. Mars was more dry, a little colder and hotter, and had its own plants and creatures, but otherwise they were much the same.

[At one time, some of the people of Mars ruled parts of Earth. At another time, some of the families of Earth ruled parts of Mars. To the ordinary people, and to Singers like me who are not political, these things made little difference. The two planets traded freely. The culture of each planet affected the other.

[Since a time beyond memory, there were cats on Mars. They were among the native races there. Their peculiar wisdom made them highly prized as companions. But they chose their own locations and never submitted to ownership by anyone.

[Some time after the pyramids were built on Earth—but long before the Great Flood that is so famous here—a number of cats decided to colonize Earth. They traveled here with people of Mars, and stayed and prospered, populating the Earth with cats. Earth cats became furry, but they looked much like Mars cats.

[Mars and Earth were almost twin planets, like the twin moons of Mars, Daimis and Febos, which in those days were the same size—smaller than the moon of Earth but just as round as it is. The two planets often worked together like one person and their Other, like lifetime friends or brothers.

[Then the wars started. Some said it was an evil Martian ruler who did the first damage to Earth. Others claimed that the first damage was done by Earth armies who used weapons against Mars.

[It's true that those who loved having the power to destroy seized the opportunity in these wars to kill and ruin. They chose to damage and they did. But although you may be shocked to hear it, I believe that the wars were started by the cats. Long years of listening have made me decide that this is so.

[One couldn't call cats political, and most are far from being evil, but in those days they were fond of intrigue, and nothing restrained them from plotting and creating secrets. They became interested in making people act and react by telling them bits of gossip about each other. It was a game, I believe.

[Like many games, this one went out of control. The cats of Mars and Earth never intended to have the effect that they did. And not all were playing at this, but many were. They made a Martian ruler worry about attack from an Earth emperor. They made an Earth monarch expect a fight with a Martian queen or with the Council of Elders. Suspicion led to acts of war. The conflicts increased in complexity.

[On both planets, too much cleverness and creativity went into making weapons and defenses. The people remembered the old songs less. They created less. Without the songs the electricity of Mars became unhealthy and the balance of water on Earth was disturbed.

[On Mars, the ancient dreams of honesty and beauty

were lost. Our most prized artists and their most splendid creations perished in the fighting that began. And on Earth the same. The ordinary people became slaves to wars.

[Then the Singers became important, on Mars, to work at restoring the balance for all people. On Earth, it was different: priests and magic were what they relied upon to heal the problem. In neither place did anything heal the problem completely.

[Only when the air was perishing on Mars, and both people and cats were dying as the cold took over, did the cats of Mars fully realize what they had done.

[It was too late. Earth had "won" the war. But really, both Earth and Mars were scarred and broken by it. Both had lost.

[Now, as you know, cats are proud creatures. For many ages the bodiless cats of Mars slept, resisting the knowledge of what they had done; they lost themselves in their ancient moments of glory, wanting to forget the rest.

[It was the Nile cats who retained the wisdom to see the truth at last. They were repentant. They called for an end to the long era of blindness for their race on both planets. And out of the death and dreams of an old era they began again.

[Freya brought this beginning to me, too, on Mars, by reminding me of my calling and my promise to my people. This work is hard to do, but only because so much was stopped still and dead, and it must now be started from death again and made to grow.

[And as you have seen, the cats are intent on repairing the damage they have done—as much as they were absorbed by the game that corrupted them in the first place. You see, now, why they have tried so relentlessly to get your help.]

She was finished.

Freya had heard it all. Was she the only one who had? The others all seemed to be asleep, lulled by the healing that Ahanith could not keep out of her voice.

Freya could feel that she wasn't offended, though; and one way or another, they had probably heard her.

**

Ozzie woke. Or had he been awake? He wasn't sure. The others were out, though, absolutely, and bunched together next to him at the table like a litter of puppies.

And there were the Singer's eyes, still: hovering nearby in the center of the galley. The galley radio chattered on in the background, and he submerged again into a dream about cats.

**

Norm had taught Freya the stones-paper-scissors game—one way to warm up your hands—and he was beating her at it. Down here in the galley the boat creaked and rocked, held snugly by the sea.

They were sailing off the dark southern coast of Iceland now, traveling westward toward Reykjavik. With Raker's demand for extradition on the news every day, it seemed to Freya that they kept moving back and forth like a chess piece, trying to avoid being taken.

The galley radio news was interrupted by Ozzie's yell from above. "Hey! Attention all hands!" he called down to them through the hatch. "We're being chased again! All hands on deck!" Norm and Freya dropped their mugs in the sink, hauled their boots on and stuffed their sweater-sleeves into the heavy jackets.

Norm got to the ladder first but she was right behind him, holding on while he climbed. The boat tilted as Ozzie took it into a fast turn to the starboard side, and then it accelerated. Freya went rigid when she heard Alexis' squeal and a thud, as if she had been thrown into a wall up there in the freezing wheelhouse.

On the galley radio behind her the chatter cut to Raker's voice, live: "Discussions have been difficult but I am happy to announce that the ring of international conspirators will be extradited by Iceland—"

She sagged. "How *could* they?" she yelled at Norm's boots, disappearing upward ahead of her. One more betrayal, on top of all the others: her own country, turning her over to that creep. She struggled against sudden heaviness, trying to haul herself up the ladder to the deck.

The noises above, closing in, made it clear that their

pursuers had more engine-power than they did. *Friends of Raker*, she thought bitterly. Ozzie must have jerked the steering hard; their boat reeled sharply to port. The motion swung her weight roughly, tearing her grip from the ladder. Then with a crash, something hit the starboard side of the boat, changing its course again like a billiard ball.

She pitched through the air onto the dark floor. She heard Norm yell and crash into something heavily above. She screamed: "Grab it, Norm, don't fall overboard!" Light from the other ship poured down the ladder toward her as if it was trying to find her. *Norm, shut the hatch...*

There was a slam from the port side, now, that sent cans crashing out of the cupboards in the galley. The impact threw the Singer's goatskin from her cupboard, across the galley into a bulkhead. With a crack, the lid that closed the mouth of the flask shattered and flew in pieces through the air. Numbly Freya knew: there were two boats out there, and one that Ozzie didn't see had just hit them.

The Singer shrieked. A sharp hiss like wind traveled through the galley into her cupboard and slammed the little door repeatedly into the bulkhead. [They always win. Always win. Always win!]

If she gives up, is there any hope?

"NO!" Ozzie bellowed furiously from above, as if he had heard Ahanith. Or were they attacking him?

The Singer was gone. The ship tilted frighteningly. Someone screamed up there.

In a fury, Freya leapt to the ladder and stormed up it. *I will fight those filthy scumbags till I die. I don't care what they do to me.*

CHAPTER THIRTY-THREE

OZZIE GRIPPED THE WHEEL, prepared for the next impact.

Out on the deck someone screamed again, hideously. It made his neck prickle. The scream paused, and then started again. *Alexis?*

"Alexis!" Freya called from the wheelhouse doorway. "Where are you!"

Ozzie couldn't stop to look. Horrifying. Was she hurt badly? The screaming went on.

The engine roared as Ozzie held on fiercely and danced a roughhouse, delicate dance with the boat. It was like flying a glider in gusty wind; maybe their fragile old boat would split from the number of gravities he was shoving it through.

They suddenly pulled out of the terrific drag of the sea—must be going 80 across the surface now, freed from the resistance of the water, the keel of the boat slapping the tops of the waves. The scream let loose again. His body shuddered with chills from the horrifying sound.

[Always win, always win, always win—]

"NO!" Ozzie bellowed furiously, again. He had a second to look now. Alexis was right beside him, sitting weakly on the wheelhouse floor with her forehead cradled in her hands. "Oh Ozzie. They always seem to!" she wailed. Freya knelt by her, checking for injuries.

"How 'bout *just one time* when they DON'T?" Ozzie bellowed back, over the noise of the engine and water. And the screaming, again.

Now the other boats—one on each side of them—peeled away in opposite directions. Something was ricocheting back and forth between them, screaming something like words, he thought, another language, the same thing over and over.

—It was singing. Sort of.

Ahanith's famous death song? Could that be it?

The voice screamed again, loud enough to split water into gases, and the words hit his mind like a gut-punch:

[Death to destroyers, destruction, confusion, chaos to chaos, destruction, confusion! Death to destroyers...]

Alexis shook her head, as if she meant to clear her ears, and rested it on her palms again. Ozzie steered.

Half a minute passed. The sea quieted gradually. So did the screaming.

"Did you see that, Freya?" he grinned incredulously at her. "The Singer scared them away, I think!" Then he saw Freya's warrior eyes and her flaring nostrils, her panting fury. She looked like that screaming song sounded. Better, of course, but just as mad.

"Where is Norm?" she asked.

He didn't know.

Dread seeped into him along with the cold, and he turned quickly back to his steering. They had casualties. Right now he was pilot, navigator, and captain.

**

Another minute passed, with no sign of their attackers, before he looked around again. Freya had disappeared. But Norm arrived in the wheelhouse doorway; Ozzie gasped with relief.

"Thanks for finally slowing down a little, man. I had to hold on to that fish-tub for dear life." He yawned. "Well, they're gone. Guess the Singer out-shouted them. It's too bad the Growing Song doesn't have that much punch! You OK, Alexis?" He patted her back.

She didn't answer. Norm shone his flashlight around at the wheelhouse and Ozzie looked where the light fell: it was sad to see the navigation stuff broken and scattered all over the floor. And the shattered window on one side. *Bastards.*

"It's broadcast time. I'm gonna broadcast now anyway," Norm said. "Show them."

The ship seemed to be off-balance. The wheelhouse floor was tilted. "Yeah," Ozzie said. "Alexis, Norm, where's Freya? Go see what it's like below, huh?"

**

The sea was milder now, closer to land. Reykjavik harbor must be just about a day along the coast from here; and it beckoned, but Freya had already begun to steer them away again, out to sea. Now that they could be extradited by Iceland, too, they needed to get far away from here or be captured.

—*Get far away? How?* They were leaking. Ozzie sighed as he landed at the bottom of the ladder. Time to send Alexis up on watch. Norm sat on the sea-slopped floor of the little galley, elbows on his upraised knees, staring at his destroyed broadcasting equipment.

"—Who are the people who are always wrecking things?" Alexis was saying. "Most people don't go around attacking people, destroying things, stopping good ideas. Do they? Who *are* these people?"

Norm went on staring.

"You know, Norm? I mean, what kind of person would—"

Norm looked up at her. He didn't seem to see her.

Sympathetic tears began to run from the corners of her eyes. She knelt beside him and put her arms around his neck, crying into his hair.

Ozzie couldn't stand it any longer. "Alexis, better get the pumps going on the life-rafts." He went above to find Freya again.

**

Freya held the wheel steady. She looked much calmer now. But her wet hair was matted to her head and her hands were grimy. She must hate that.

"You OK to go a little longer?" he asked.

"Yeah," she said. They were headed out to sea, southeastward. The lights of some small Icelandic coastal city were receding, far behind them. She turned to him and nodded a little. "It's good to drive this thing," she said. "Makes you feel like you can make something do what you want."

"Yeah." In a minute he said, "You're a good pilot, a good navigator. You could be a space trader."

She looked sideways at him and sniffed, but she smiled.

"What's your plan, Miz Navigator?"

"Get as close to Norway as possible before the poor boat sinks, and then get in the life rafts. You agree?"

"Yeah. Long way to Norway. And cold, but we can't stay here, anyway. That seems best."

After another minute she said, "Think we'll win, Ozzie?"

"Dunno."—*Are you kidding? Not likely.*

"Know what I think?" It sounded like a brand-new thought, one of those things she just thought of, out of the blue, all the time: "Evil people deserve to be beaten. Someone has to do it. Why not us?

"—And look, I think they will help us beat them. I think the guide was right, in Egypt. They will help us beat them. Maybe we just need to laugh at them enough." She looked to see if he was still listening. "I think we should treat this like a killer game of volleyball—just go like hell and win or lose but have a good time anyway. What do you say?"

Her face was utterly brilliant and glamorous, like the one that launched a thousand ships, and she smiled a delighted smile at him.

Her brilliance bathed him in crazy bravery. He felt like a thousand ships, about to be launched. He thought of Malo and the gypsies, grinning at their opponents.

In that smile and in those eyes, I can do anything, he thought. What the hell. Although it took effort, he grinned back at her. "Let's do it," he said. But he realized she had already heard him.

**

They left the boat cruising slowly for the moment, now that they were in the open ocean. It was time for them to warm up and send the other two on deck.

Ozzie followed Freya down the ladder to find Norm idly picking up broken parts. Alexis was helping him, arraying hers neatly on the little dining table. Freya shook out her hair and peeled off her coat, so Alexis dutifully began to put hers on.

"I hate losing," Ozzie said, looking at the messy

electronics.

Norm looked up and shrugged. "Everybody loses sometime or other."

"Yeah. Well, I hate losing to assholes."

Norm chortled. He looked Ozzie in the eye.

Ozzie grinned.

"Right. Now that's a special case." Norm slapped another part onto the table.

He continued to pick up wet pieces for another 30 seconds and slap them loudly onto the table. Ozzie and Freya helped.

Norm stopped and looked around at them all. "You're right, Ozzie. Let's beat the tar out of those assholes, buddies."

**

Alexis went up on deck and took the wheel so Norm could put the station back together. With Ozzie's help to dry the pieces at the stove and organize them, and Freya's help here and there on the little fussy parts, he got his broadcasting rig reassembled pretty quickly and most of it worked. Most of it.

Where the parts had water damage he now worked fast to find the problems in those, too. He said, "I'm going to deliver the next broadcast anyway, even if we sink while I do it."

As if to oblige him, the ship tilted a little more. Alexis had them going as fast as possible on course toward

Norway, but the ship was foundering a bit already. "Life rafts pumped full yet, Alexis?" Ozzie called up to her.

"Aye, sir," she called down, and saluted. "Both pumped full and lashed to deck."

"...One damaged part that I have to replace," Norm muttered.

Freya put the galley radio by the stove to dry and while it did she packed up most of the remaining food, some for each raft. And some water.

"All four backpacks packed, fellas," Freya said. "What did you do with all your clothes, Norm? Not much there."

"Gave them away," he mumbled. "To make room for players. Damn. I just need one thing... One part missing to broadcast, Oz. Thought I had a spare... Hey, check my phone, Ozzie, willya? I think I got a photo of the turkey-shits. One or both boats, captured in the act."

**

Freya had no idea what time it was. Ozzie had put a few cans of things together into a skillet. Dumb as his dinner looked, it smelled good to her. Norm was still hunting for the missing part. They had to get out of here; the water on the galley floor was all over on one side now. And there was more of it all the time.

"Ahanith was banging the cupboard?" Alexis repeated, receiving a mug of hot tea from Ozzie.

"Yeah," Freya said. She cupped her hands around her mug. "If you don't believe me look at the dents in the wood. It hit the handle on the opposite side, it banged so hard."

"That's a good sign. That she can do more things, I mean." Ozzie drank a scalding mouthful.

"Never thought of that."

We shouldn't be talking about her as if she isn't here, Freya thought. But she didn't seem to be. No sign of her for hours.

[Ahanith?]

A long silence followed, in which they ate fast. The ship groaned and shifted again, leaning a little lower on the port side. Then Freya heard a rushing sound.

A pair of turquoise-green eyes showed up inside a mist, and two hands.

[That was you who scared the ships away, wasn't it. You were fierce!] Freya said. The Singer's anger seemed to be fading.

[My friends.] She waved her long fingers expressively. [I am sorry.]

Freya gaped. [Sorry? Why?]

[I have failed to fix the volcanoes. I am not strong enough for what you need.] Anger started to build around her again.

"That's silly," Ozzie said out loud at the almond-shaped eyes. "You were strong enough to make two boats full of nasty bastards run away! Besides, no one has failed. Not yet."

[Not strong enough yet, either,] she repeated. [But I

will not give up, my friends.] She began to sing softly, as if to herself, and drifted off toward the bow.

Norm stared at her. "Oh!" He stood up abruptly. "I left it in the head with my toothbrush, so I wouldn't..." He disappeared and returned with a small piece of silvery metal. "The spare part! Found it—fell on the floor. Thank goo'ness! Help me, willya, Freya, hold that end and I'll..."

While he tested the result Freya checked her phone for newsfeeds. No phone service, of course.

She took the old ship's radio from the place where it was drying out, next to the stove, and tried turning it on. It sputtered and picked up signals. Some kind of news chattered on, but there seemed to be nothing about volcanoes. At a hand signal from Norm she shut it off again.

Norm switched on his equipment and began, in his pretty-famous radio voice: "Apologies to our millions of devoted fans worldwide (he coughed politely) for the interrupted MuckRaker Radio broadcast schedule. What interrupted us? An attack on our seafaring radio station by two motor vessels at least three times our size, with obvious intention to sink this boat.

"And the attacking ships go by these names, in case you see them and want to let them know what you think: *White Knight13* and *Good King Wenceslas*, registered British motor vessels. Thank you to our research department for the fact-finding...

"Water's leaking into the station here, but instead of

sinking we have Saved Our Ship, fans, at least temporarily. So our message for tonight is SOS! And for now (until we sink) we are on the air again! You heard that right, S-O-S! Tonight we'll be playing until we sink, vintage oldies and all the best rock.

"Beginning with a song that you will want to join in on..."

And he beckoned Ahanith to the mike.

**

The boat was leaning even farther to port now. Norm had the station robotics queueing up the next hour or two of music for automatic play. *Is he crazy?* Another groan from the wooden frame of the boat, and the dinner plates, set out to dry, slid along the tiny counter.

Ozzie was clearly thinking about when they should pull the plug on the broadcasting and get into the lifeboats. *It better be soon,* Freya thought.

She turned the old boat radio on again, just for something new to do for a minute. It sputtered a little, then:

"Tonight," it said, "the Icelandic government formally refused the demand of U.S. Representative B. Arnold Raker that Iceland extradite a mysterious student and an Icelandic firefighter..."

"Listen!" she demanded, searching their faces.

"What?" Alexis' head appeared up in the hatchway.

Freya realized they couldn't understand; the voice

spoke Icelandic. "It says, extradition failed! Icelandic President says—" Her voice caught in her throat but she looked at Ozzie's eyes and remade her face into a fierce gypsy smile, full of teeth, "—he says, 'We will not give up our citizens to other nations for trial. If we decide they are to be tried they will be tried here in Iceland.'"

She turned the tears that threatened her into a wild yell. The others howled with her.

"Hope they told Raker what to do with it, too!" Norm said. They howled again.

The noise didn't drive Ahanith away; she just moved back and forth in the room as if she were pacing. Her eyes brightened and dimmed a little, pulsing like embers, as she sang.

The boat bucked on an irregular wave. The dinner plates slid from the counter and clattered to the galley floor.

"Say goodnight and pull the plug, Norm!" Ozzie said.

Norm didn't budge. His attention was riveted on a tiny display on the cube of broadcast equipment, where numbers showed broadcast volume or something.

"—Well, you can go down with the ship if you want, but I'm not. Freya, I'll hand the packs and food up. Go up and catch, OK?"

"But Norm!" Alexis called down, still kneeling above the hatch. "Aren't you going to send off the contest package? We have it all bundled up to go! If we sink, we may never…"

"What contest?" Ozzie asked. "Are you losing your minds? We need to get into the rafts."

Norm grinned up at Alexis. "I forgot!" His grin became evil. "Yes, let's drop the contest bomb on an unsuspecting world." Alexis slid down the ladder while Freya shouldered up it.

As they hefted packs, Freya and Ozzie stopped long enough to watch: Alexis entered all the settings on Norm's phone and the Mystery DJ cackled as he pushed the button. "Bavoom!" he provided the explosive noise himself. She giggled.

Then he interrupted the station's music and announced the contest, rules and cash prizes and all, while the packs were being boosted and tugged up the ladder to the life-rafts. For good measure he repeated the SOS message one more time.

**

There was a foot of water on the galley floor now. With Alexis holding the wheel, Freya called down the ladder: "Quick! Let's get off this boat!" The rafts were loaded with food and water up there, a bag in each, and the packs, two in each. Ahanith, in the goatskin with no lid, was tied to Freya's pack.

He and Norm struggled to position a box filled with electronics so one person could carry it up the ladder.

"It weighs too much to put all this junk in a raft, Norm!"

"But we need it."

"Not if it sinks your raft. You won't need it at all then, will you?"

Freya clattered down to help push the box upward from beneath. "Move, friends! No time to waste!"

"Hey!" Alexis called down from the deck. "Hurry, quick! A boat off to starboard! Icelandic. Think they want to give us a tow?"

"Kidnappers! Tell them *no way*, Alexis—"

"Hush up a minute, Norm," Ozzie growled. *Even if they **are** kidnappers, how could we escape from them now? We are wallowing.* Time for some diplomatic discussions. "Here, take the box. Freya, quick, come translate!"

They left Norm standing by the electronics box and scrambled up the ladder.

CHAPTER THIRTY-FOUR

"WE HAVE WON THIS ROUND!" Freya declared.
Ozzie noticed how pink her cheeks were from the cold, and how messy her hair was. He liked it that way. They were *all* messy, and fishy. And safe, for the moment. *And if "won this round" means "we have survived it," then yeah.*

"Yeah," Norm yawned affably, looking around them at the inside of the rescuing boat. "I thought these guys were just more boat-bullies, but they're really OK." *Hringur* was being tugged, limping, behind them, like the wounded after a battle. "Did you hear what the guy said? They picked up my SOS broadcast. They were searching around for MuckRaker Radio!"

Alexis' sideways smile went up her face. She was the perfect person for Norm; she almost always thought he was funny.

They had a damaged boat to fix, but at least they weren't going to be extradited. Maybe they could sell their load of fish for enough money to fix the boat. They weren't sitting on the waves in lifeboats right now. All told, things were looking up.

Ahanith had just appeared before them as a pair of green eyes, back out of her goatskin already. She didn't even seem to be seeking a cupboard. She stood people-sized, almost, taking in the new environment just as they were: another galley, bigger and less clean than theirs. Lots of dark wood. Dim light. Warmth.

When they had let their backpacks fall to the floor and dropped into seats at the galley table, Ozzie looked at her again. She stood still, as if she were listening to something he couldn't hear.

Their benefactors brought them hot tea or coffee to drink. Ahanith inhaled the minty steam from Freya's tea. After a few minutes she told them dreamily:

[Malo sends greetings, glad that you are safe. They sing with 180 around the fire these days. Diana has been recording. It is a cold winter but fires and singing make everyone warm. Malo thanks Norm for distributing the players. They have all traveled to their destinations, north and south, and the music has gone still further.]

It was clear that she and Malo had a hotline of some sort between them, at least some of the time. Ozzie hadn't thought of that. He said "Norm, maybe Malo didn't need a phone to talk to you and Ahanith."

[Ahanith,] Norm said. [Please remind Malo that Alexis and I sent String and a cat named Tom to the

London campground and northward from there through England, carrying players and slivers for the campgrounds to copy.]

The green eyes got intense and beautiful and she was silent again for a while.

[Malo, who knows String best, has looked and found her, Norm. Far away...] Her eyes glowed brighter, then she continued, [String and Tom are on the steppes of Asia, traveling toward Singapore, Hong Kong, Tokyo? or such places along the water. They have caused distribution of 5000 slivers of sound: to campgrounds of Paris, Berlin, Milan, Budapest, St. Petersburg, Istanbul, Jerusalem...]

"Alexis, look what we started!" Norm half-stood and bowed.

"They really took the bit in their teeth and ran, too," Ozzie laughed.

[...Tom found the Nile cats, who were willing to take the players and songs south to all the campgrounds of Africa...Then he and String went to Nepal and Tibet, New Delhi...there they caused 1000 more, String says...]

"Outstanding!" Freya crowed.

[...And String said something complimentary, Norm.]

"Yeah, I think she likes me." He took off his glasses and polished them on his sweater-front. "What'd she say?"

[She said: "He is learning to pay attention and did some intelligent things."]

Freya looked around at them and said, "Well, those two cats are being heroes. And what about us: these attacks have kept us running and hiding and defending ourselves too long! It's time for us to fight back, instead of running."

"Sounds good to me," Norm yawned. "But who exactly do we go after? Besides Raker I mean." He grinned his deliberately annoying grin.

"There's so much we still don't know…" Alexis said.

"But we know enough, don't you think?"

Ozzie wished that they did.

"We could start with a court case against Raker for trying to destroy our boats in the North Sea," Norm said. "As soon as we can prove it. I'm gonna find out what we can do about that. Maybe call the North Sea Fisherman's Better Business Bureau. I wonder if it would be bad karma to do that."

"Maybe *Raker* has bad karma, Norm, and you're it," Ozzie grinned.

**

They were on their way into Reykjavik harbor, and texts started coming in on their phones.

"Damn phone," Ozzie said, eagerly looking. One message from Diana had been delayed: a PII document showing that they all had been hired for months as researchers for PII in Britain, Iceland, etc. "Look: Norm and Alexis, you're employees."

Norm took a look. "At pretty bad pay," he grinned obnoxiously. "We've made more money fishing."

Another from Diana a couple of days ago updated them on the latest PII measurements of the volcanoes. Nothing remarkable. "Hey, here's more," he said.

> **Our new private investigator has found out that the Oklahoma "testing" company is owned by another company, with names of people on the board of directors that the investigator has not been able to locate anywhere. Yet.**

"And some more from later:

> **PII's geo team has completed the survey of vibrations, over the entire surface of the planet. There are eleven more "testing" locations like the Oklahoma one, all sending out the same vibration: the one in the ring formula.**
>
> **Twelve locations total. All just in rural areas, no particular place, on all seven continents. Under the ice in Antarctica, buried and camouflaged. Out in the bush in Australia, on the Gobi desert in China. Whoever put them there must have thought they would never be noticed or located—at least until it was too late.**
>
> **But the company that owns the Oklahoma testing company has no connection with the**

companies that own the other "test facilities." The investigator has done extensive checking on this.

"Whaat??" Alexis said. She shook her head at Freya. "I don't believe that." Freya didn't either.

"Wait," Norm said. "I just thought of something. About that message Malo sent us about the cats, String and Tom: The slivers are all over Asia now. But how could the people in all those places play our slivers? It's not too hard to copy slivers somewhere, even at a fuel station. But the right players? Not everyone would have them lying around."

Alexis wrinkled her nose at Norm. "So some of those people holding slivers have no way to play them, really."

"Probably *many* of them."

Freya groaned inside. More roadblocks. "We'd better fix this. Could Malo help?"

Norm called. It was night there but they got him and live-holoed him in. A small image of him stood on the center of the table, looking warily around, listening intently to the boat-horns out in the harbor but smiling brilliantly. Was it singing that Freya heard coming from somewhere around him?

Norm told Malo about sending the song to radio stations, as a contest. Malo nodded, looking pleased. Good news first, Freya guessed.

Norm congratulated Malo on String's Asian trek and

her triumph: 5000 slivers out, at least. But then he explained the problem about the players—how it was that many of the people with String's slivers would have no real way to play them.

"We should send the song to all of them so they can sing," Malo said simply.

They were all silent for a minute.

Alexis said, "Speaking of sending, here's an idea. One of those singing telegram holosheets. They're a big deal in London this year. One that sings as soon as you open it up."

"Yeah!" Freya said. "With an introduction recorded from Malo so they believe they should sing the song."

Alexis said, "We could send the telegrams out as a blast to local distributors who get them delivered…"

"But those telegram sheets cost a lot," Ozzie said. "To get thousands out…How would we pay for them?"

"I think the company (PII) should pay," Norm said. "—Well, Ozzie, Diana says we're working for her now, right?"

"Jeez, Norm. Not sure they have the money, either."

"Just saying."

Freya looked at Alexis, who was punching queries off to the worldweb and watching them come back in:

"Quantity discount price for 100, $16 U.S. each.."

"Sixteen hundred dollars…" Ozzie said.

"Discount price for 500 is $5 U.S. each…only $2500!

"Only?" Ozzie said.

"...with worldweb business in singing messages at a going price of $25 US dollars, we could make almost $20 profit on each, so we'd only have to sell 100 singing telegrams to pay for 400 free ones!"

The others stared at her.

"Who could afford to buy them from us?"

"Who will sing them?"

Alexis looked up into the air for a couple of seconds.

"We can sell special 'happy birthday' telegrams for *twice* that much, as fundraisers, to maybe 1500 rich Oxbridge alums who want to send the 'Happy Birthday to You' song to someone, with the extra $25 on each as fundraiser funds—"

"But who will sing them?" Norm demanded again.

"Oh, Ahanith will I hope. With the Oxbridge Chorus, of course! And they'll get on the phone and sell them, too. It will be their fundraiser!"

And Ozzie thought he was a good trader. Freya grinned at him till he laughed.

"But wait," he said. "How many languages will all these people speak? There are thousands of languages in Asia. Wouldn't we need to have Malo's message translated for all of them?"

Malo said, "I can take care of that."

CHAPTER THIRTY-FIVE

BEHIND THE LARGER FISHING VESSEL, the wounded boat *Hringur* wallowed into Reykjavik harbor in gray light. Whether it was dawn or dusk or both, Ozzie didn't know until he looked at his phone. Icelandic days still confused him. But one thing was certain: this was a safe harbor for them, now that Iceland had refused Raker's demand to extradite them.

To their benefactor, the captain of the rescuing boat, they gave their grateful thanks and offered whatever was extra in their cargo of icy fish. They would keep just enough to sell for repairs, Ozzie thought, and some to thaw for cooking later.

The towing boat still had its winch connected to theirs. In return thanks for all the fish, the owner further insisted on winching up the damaged bow of the *Hringur* so it would stop taking in water and the hull could be pumped dry. The bow was caved in a little and punctured here and there.

So Ozzie and Norm, Freya and Alexis stood on the vibrating wharf next to their rafts and packs and the box that contained Norm's broadcast station, watching their boat rise upward one creak at a time. The life-rafts sighed softly beside them as the air sagged out of them.

"What a mess," Alexis shook her head.

Freya hadn't stopped smiling for hours, it seemed.

"You still happy because Iceland talked back to Raker?" Norm goaded her.

She scoffed. "More people need to talk back to him," she said. "I'm just happy when something fair happens somewhere."

Boats and ships came and went busily in the harbor. While Freya got a message off to Doug and Ilse, Ozzie texted Diana.

"Norman Garcia, right?"

Ozzie turned, just behind Norm. The guy who had sold them the boat stood looking at them. Ozzie felt his face getting hot: the man would think they had been so careless or stupid that they had wrecked his boat already.

They shook his hand politely. Ozzie hoped the guy was a hardy enough fisherman that he didn't mind getting a lot of fish-smells with the handshakes.

"I see you've had trouble."

Ozzie explained.

"Well, I would like my boat back. Your ring was not good... defective."

Ozzie blinked.

"It gets hot sometimes. And no one can cut the band to resize it."

Ozzie stared.

"*Ja*. True. There is some kind of cheap metal inside, too hard to cut!"

Cheap metal's not hard to cut. Ozzie frowned. "Show me."

The man pulled it out of a jacket pocket. Someone had tried to cut through the thick band and had only managed to expose a thin dark wire at the center of the brilliant gold.

Norm held out his hand to see it too. He looked at the ring, the black center, and at Ozzie. "Hmm," he said.

Ozzie accepted the ring back and they signed some papers. Norm signed a note to pay for the damage to the boat and they let the man split their fish cargo with the rescuers to cover the 10% they owed him for the month. It was the best deal they could make. "We'll all help you pay it off, Norm," Ozzie said needlessly. But how long would it take to do that? He'd have to make a trade, or do something clever.

No time to think about it now. Now, the only question that mattered was *what is in the ring?*

He texted Diana again.

Norm said, "Now that we aren't gonna be extradited, boys and girls, we can be landlubbers again, right Freya? Well, then, we have things to do. Let's get all this stuff to your house." He yawned and stretched. "I need to get the radio station reopened quick. Think your mom will mind?"

**

Freya knew Ozzie liked his dad but she had never seen them so happy to be together. Doug's weathered face still looked suntanned, even after months away from New Mexico. His very-straight nose and hard jaw still looked stubborn, like Ozzie's, but his patient eyes were brighter now than before.

Doug and Ilse were on the doorstep when the four of them arrived at Freya's house in a truck they'd rented on the wharf with their last fish cash. Her mother's eyes crinkled with merriment, as usual. The house vibrated without stopping right now, but the kitchen was daylit for the moment and cheerful and it smelled like fresh coffee.

"We heard your station every day, Norm! Liked your music.—Did you know? We decided to do the song too. With Icelandic words. It's popular here. And we have an idea that might help—"

"Hey, you have to show us your amplification system. Maybe you're getting better ratings from Diana than I am because of your amplifiers. Do you have them set up here where I can see?"

When Ahanith entered the room, appearing as green eyes and a sweep of white, awe made Ilse rise to her feet to greet her. Before long, though, they were singing together quietly like old friends, nearby in the little living room. Doug joined them. Then Freya and Ozzie.

Freya couldn't remember when she had last slept. So many watches, nights and days: confusing. But the music lulled her. It was still gray daylight outside when she fell asleep in an armchair listening to her mother and Doug and Ahanith sing.

**

Within hours Norm had the station all set up in the basement, Freya's room. Ilse and Doug's instruments and speakers had been cleaned out of the extra bedroom upstairs because Diana was due to arrive, her mother said. So Freya's thick-walled, cool retreat now contained a pile of musical gear and another pile of backpacks and sleeping gear; and an old table had been added as the new location for Norm's entire radio station.

"Beats having it on a toaster shelf," he said. He flipped the switch and pushed the button for the echo-effect. "This is Radio Reykjavik, son of MuckRaker Radio, now operating from a secret land location," Norm intoned in his most resonant voice. "Bringing you the finest vintage sounds, cherry-picked rock, and underground info from those who know!

"Speaking of info, do you know who hired those two ships that almost sank our broadcasting boat? United States Representative B. Arnold Raker, and thanks again to our research department for finding that out!

"As always, we begin our broadcast day with the Growing Song... mandatory singing time for all Icelanders! Come on, Brits, help those Icelanders beat the volcanoes, sing the bloomin' song!"

**

Ozzie said quietly to Dad: "I thought that when the school year ends for Ilse you two might go back to Las Cruces. Safer there..."

"Safer? Who knows. But maybe safe isn't what we want. Ozzie, there might be some adventure left in us, old as we are."

Ozzie flushed. "I didn't mean—"

"Anyway, the band is on its way here. To Iceland."

He grinned at Ozzie's shock.

"We have an idea, see."

**

"Here's the plan," Dad finally said, after Ilse returned from her afternoon at the school and they had all gathered to listen. The kitchen was brightly lit, shaking back and forth about the width of Ozzie's little finger, and steamy from coffee and tea. The darkness looked safer where it was: outside the windows.

"During the last month or so Ilse and I have sent messages to all our musician friends to ask them to sing the Singer's song some way or other. The singing will at least make people aware of the problem, until a real solution is found. Right?

"We've put out notices and holovids all over the worldweb to get other musicians to record and perform our version of the Singer's song. But not just that. To write their own songs that include her song, and also come up with their own ways of singing her song raw—"

"You could enter my contest!" Norm grinned. "And so could all your friends—"

"—So send me the plan and I'll get it to them."

Alexis wasted no time in getting a used pulp puzzle-book, her scrap paper resource, and began scribbling down addresses for bands, whatever Ilse could give her. The addresses of the notice sites too. There were a lot of them, and they huddled to list them out while Doug went on.

"—Anyway, you probably haven't had the time to look around on the worldweb much lately, but our holo-posts went planet-wide. People are making this music happen all over the place—live, on the web, by holovid channel, even on some old video channels and using some pretty primitive soundbite-relays in some places..."

Ilse smiled delightedly around at them: "You should see all the bookings for performances of the song—"

"—Right," Doug said, "hundreds of them now, in stadiums, rock venues, all over earth, a lot of them starting at New Year's Eve—"

"One week," Freya said, and her eyes began to shoot excited sparks.

"That's right. They'll be gathering at loads of venues to play and sing this music.

"And now we have permission from the Reykjavik City Council to host a music festival right here outside the city, near the Hengill volcano site, also starting on New Year's Eve. And lasting as long as we want."

Hengill. Ozzie saw Freya go still. All the joy went out of her face.

"No kidding, Ozzie. Ilse posted it a few days ago. We've already received more than a hundred entry requests from bands."

Norm launched eagerly into talk with Doug about broadcasting from the festival site and wanting his station to do the lead coverage for it. Ozzie had never seen him so outright excited about anything.

Ozzie tried to imagine Hengill, crouching outside Reykjavik as a cold memorial of its own past eruptions. Freya had told him about it somewhere on the road to New York harbor. After hundreds or thousands of years asleep, or mostly asleep, now it had begun to come alive.

Had the others ever seen holovids of volcanoes close up?

Of all of us sitting here, only Freya knows how scary the volcanic fires are, live.

**

They all went to the Vestur thermal spring and public pool for showers and swims that night. One bathroom in their little house, and 6 people: this was Ilse's solution.

"I called you once when you were here, didn't I?" Ozzie asked Freya, recalling the name. Ozzie took in the cement decks, the brightly-painted supports and beams of the roofing over the pools, and the steaming water. He breathed the sulfur smell, experimentally.

We didn't know each other. And now our lives are so mixed together we couldn't possibly be strangers ever again. A big change. He guessed she was his girlfriend, without any ring or bracelet, but there was hardly any time to act that way. The way they were was more like a promise for a future; that was all. He kissed her on the nose, feeling lucky anyway right now.

They climbed in and soaked in the hot water at the edge of the pool, resting their chins on their forearms, and talking softly together while Norm clowned and Alexis giggled at the other end.

While they talked he realized he was staring at the wrinkles on the surface of the pool between them. The surface quivered so much that the light it reflected was shattered, flying fifty ways at once. *The earth is shaking harder and harder under us.* The wrinkles were a menacing message from somewhere deep in the earth.

Suddenly the message made everything that was important to him seem weak and helpless. *This pool could be a geyser tomorrow, and the street out there could be a river of lava. The big changes in our lives, which are big events in most lives, are nothing compared to what could happen any time now...*

He must be getting used to feeling doomed, because now he kept on thinking in spite of it: *Just need to find the right things to fix. What are the right things? Are we anywhere close?*

"Just a few days till the Reykjavik Music Festival starts," Ozzie said, instead. "After your description of Hengill I thought they must be crazy. But my dad said they picked a site that seemed safe—"

She sniffed.

"Well, safer. Than other places in Iceland. And it's on the side of the peak, where the land folds to make a natural—"

"Amphitheater? Yeah, I know the place."

She didn't look like she felt a whole lot safer, knowing that.

**

"...but I'll never complain about inconvenience again," Norm said the next morning. "This inconvenient office is so much better than a sinking ship or even frozen fishy hands."

Alexis nodded knowingly.

Their "office" was the floor of Freya's basement room, on the old rug there, where they sat cross-legged punching the new addresses into Norm's phone and

Alexis' notebook for their next electronic blasts: the special contest blast to Doug and Ilse's musician friends, and the singing holo-message sales pitch to Oxbridge fans.

They read the musicians' addresses off to each other and cross-checked each one against Alexis' scribbled list. While they did, Freya and Ozzie were temporarily occupied with consolidating the mess in the room by stacking and shelving things.

He lifted his father's guitar case, remembering it sitting silently in their living room in New Mexico, gathering dust. *Another big change.*

Right now Ozzie could hear Ahanith sing a little, softly, as she moved about the room looking the place over. Sometimes she disappeared for a while and then returned, but he guessed that meant she had gone to visit or join the singing upstairs.

The radio station was running under robotic control right now, playing the usual vintage rock with a break for Ahanith's song at least once an hour during the daytime. Between live performances the Singer didn't need to rest so much these days. She wandered around the house singing. Or disappeared and returned singing, after visiting the local sights.—Where she probably also sang.

"How does that work, with the telegrams, again?" Ozzie asked them. "One message to the choir, and another message—"

"That's right. We'll have singing messages that we'll pay the holo-message service to send out across Eastern Europe and Asia and beyond. Malo has recorded his letter to the families and camps in a sort of universal gypsy-language or something," Norm said. "That message is coming from him. We just add what's been traveling on the slivers: the recording of the song itself."

Alexis' face was just brimming with genius. "And we pay for it by doing 'happy birthday' greetings sung by the Singer, calling it 'Mars magnetic music,' recorded on top of the Oxbridge Chorus and Choir singing backup—and I have a friend there who's getting that recorded this week—to raise funds for the Oxbridge Choir Fund, very popular—"

"—And the production fee from each Birthday-gram goes to us!" Norm crowed. "Enough to pay for delivery of the Growing Song to the rest of the gypsy world in Asia."

**

Alexis sat back and sighed. All four of them had spent most of the day in the dim basement, working on the get-out-the-info projects.

"Yes, we're doing all this," Alexis said to Freya, "making lots happen! Getting good things going. But it's bizarre that we still don't know the exact source of the problem so we don't really know what to do to fix it."

Norm didn't seem to care. He was busy showing Ozzie his broadcast-rig improvements.

Freya gloomed, "I'm sick of being stopped by what we don't know."

"...Well, here's what we learned about sound," Ozzie answered Norm. "We found some experiments, done during the last 30 years, on qualities of sound. They were based on tests from just before the Wars, showing that there's water and there's water: they found differences between one bottle of pure water and another bottle of pure water that didn't show up until they created the new tests."

"So it could be that there are different kinds of sound, too," Alexis said. "—Well, don't those PII graphs already show there are? PII tests show that one kind of sound is more effective than another. It's just that there are no instruments to measure what exactly the difference is."

"Anyway, we got that far and then ended up in the fishing business." Ozzie shrugged.

"Yeah." Freya felt a thought dawning and tried to voice it. "Same as in Egypt: we were so busy running from one thing to the next and getting each others' backs that we didn't have much time to figure things out..."

"We do have more help now, anyway," Alexis said soothingly.

"Look at this crap," Ozzie said, holding out his phone. "Newsfeed, Norm, did you see? Someone got the World Council to order Iceland to shut down Radio Reykjavik because it's unlicensed—'unfair competition with Icelandic and British stations.' Icelandic legislature promises to investigate..."

"Wonder whose idea that was," Alexis said to Freya.

Norm yawned. He went on poking slivers into slots so his radio-station electronics would be set to go for the next 23 hours.

"Well," Ozzie said, "Freya's right: whatever happens, we'd better just attack. Don't defend anymore, just attack."

"Right, but attack *what*?" Alexis said.

CHAPTER THIRTY-SIX

FREYA OFFERED TO MAKE TEA FOR THEM ALL: mint with licorice because everyone seemed to like the smell. But first she started some food. It was just the four of them for supper, so someone had to do it, and she wanted time to think while it was quiet: while Ozzie and Norm walked and talked out in the air.

She thought about what she knew and what she didn't. Funny subject. "Know" could mean that you could prove something. Or it could mean that you just knew.

One thing she did know, without proof: that Raker was hiding something he didn't want her and Ozzie to find out. Why else was an "important" person like him chasing Ozzie Reed and Freya Ilsesdottir around the world trying to stop them?

So what was Raker doing that he was hiding? *We could say that we don't know enough, or we could decide that we do.*

She decided to act like she knew some things:

First, because Raker's Oklahoma company was the next-door neighbor of the "testing" company, probably he was more than just accidentally there. Raker must be involved.

Second, the destructive vibration wasn't just the clumsiness or carelessness of someone searching greedily for oil. It wasn't because of oil or making money at all, or the "testing" would have stopped after the tests found no new oil for three years.

Third, there **was** some big important purpose for these crazy vibrations. *Raker doesn't want anyone to know who is doing this destructive thing, or why, but he also doesn't want it to stop. He seems to want desperately to keep it going, even when there is great cost and risk. For what purpose?*

The meal was nearly ready: easy omelettes that were a little like the ones Ozzie made, with all the almost-New Mexico flavors she could find in the spice cabinet. She set the tea aside to steep in the big fat pot.

Something scratched at the kitchen window. A cat sat outside on the sill in the dusky, freezing light: a big tortoiseshell creature, with golden eyes.

She let it in.

[Greetings. I'm Freya. Anything you would like to say?] she asked.

**

Alexis drifted into the kitchen, smelling the food.

Ozzie and Norm brought the cold air in with them, but they shut the door fast and hung up their scarves. "Well, we *do* need a new cat, I guess," Norm said. He rolled his eyes comically. The cat eyed him back.

As she ate Freya thought on about the destructive vibrations. *Really, for what purpose could Raker want to keep these insane vibrations going?*

The cat seemed to be very hungry. Freya handed down a plateful of leftover bits of egg, which he devoured. Then he dug his claws into a rag rug at the kitchen door, and fixed his eyes on hers as he tore at it, sharpening his claws, purring.

What purpose does Raker always seem to have? Power.

And what he's been doing to defeat us: cut us off from the people who would believe in us and give us power. Keep us running and threatened so we have no power. He's trying to take away our power and own it all himself.

"Hey, everyone, I get it now." She said it without thinking.

They looked at her.

"Look," she said. "Alexis asked, 'what do we attack?' We don't have lots of evidence, but we know enough to guess who's involved in most of the trouble. That also gives us something we can do, that we have to do: we can attack Raker by using the same trick he's been using on us, by cutting his power sources."

"Like, how exactly?" Norm did her the favor of not

grinning at her.

"I'll show you, if you agree to help me when I see the right time."

**

The golden-eyed cat sat on a chair looking at her. Its whiskers shook steadily along with the rest of the house.

The rising sun cast pale light on the counters she scrubbed. The others had left. Earlier this morning she had actually volunteered for cleanup duty while Ilse, Norm, Ozzie and Doug went to the Festival site to see the place and how the stage and sound system were going together.

Her mother's small pots of herbs rattled without stopping. She bit her lip. The shaking wasn't becoming ordinary to her. She wasn't learning to accept it. She hated it.

The local newspaper lying on the table carried the headlines:

New Volcanoes Erupt! And that was in the Pacific islands. With lava running now from the San Andreas Fault in California.

World Council Forbids Reykjavik Festival Due to Risks. That was Raker again, of course. Of course it was dangerous to hold it here. But what part of Earth was safe these days? Would he really have the power to stop the Festival?

And nothing they had done was stopping the

volcanoes, either.

Before the others had left, the early morning newsfeeds carried those same stories. "Why's Raker so against some singing?" Norm said. "He's a total party-pooper but this is pushing it too far."

She picked up the paper, looked, and put it down again. The stories all said that the Icelandic government had protested the demand of the World Council. In the U.S., the state of Oklahoma had also sent its objections to the World Council. The City of San Francisco had sent its own delegation to protest.

Where did Raker get his power? He didn't seem to know how to do much but talk. *He probably can't fight a fire, or sail a boat, or catch fish or even ride a horse. He doesn't really have much power himself. But he can just call up these people at the Council and they'll listen to him, not us. The news services, and the people who believe the news services listen to him, not us.*

That's the only power this important man, Raker, has.
"Alexis!"

**

They sat on the basement rug together. Freya's idea just needed a way to carry it out.

"Well, we have this great list of all the radio stations and media outlets," Alexis said. "Why don't we use it?"

Of course. That was the way to do it. "I hate to stoop to his level—"

"But if you hit back at someone who's beating you up, isn't that self-defense?"

Freya remembered the fierce smile of the gypsy leader who said, "When we have to, we stop them."

"Yeah, I think this is like that. Maybe I'll feel better about it if we can make it funny."

She helped Alexis write the news story. It said that Raker was a major investor in the Oklahoma fracking facility that was being sued for damages to property and farms there. And because Raker's hired boats were the ones that rammed and almost sunk a small fishing boat in the North Sea, he was going to be sued in Icelandic courts for mayhem and attempted murder. That became the title of the story that they sent: "Raker Murder Charge."

The story was nothing but assorted bad news about him; just the sort of thing he had been doing to get at them. They gave it all the glaring color they could, and somehow made him sound ridiculous too.

They sent the story out as holo-mails to all the holo services and radio stations, with an unflattering holovid of Raker spluttering while he spoke to the Council. Next they spent time looking up the news services and got it out to every one of those they could quickly make into a mailing list.

Then they waited. Freya was too distracted to read so she made a cake. *Is this the real reason why people bake?* she wondered.

An hour, then two hours ticked by before the story started showing up—on little news services first, especially ones in Oklahoma, Iceland, Malta, Costa Rica—and other places that the World Council was trying to take over, like Greece and Seychelles. Then the bigger news services began to pick up the story because they couldn't be caught not carrying a hot story like this. Soon it was all over all the newsfeeds.

"You rock!" Norm and Ozzie sent them a text from the Festival site. "Look at the holovids out on the web now. Raker's gone ballistic." Ozzie attached his own ridiculous holovid of Norm dancing wildly and grinning.

**

"First contest entries are in!" Norm proclaimed. "The winner in Bali is a group called—don't know how to say it—singing the Growing Song *a capella*, whatever that is! Sounds pretty freaky. In a good way. I'm gonna get that on the station right away, to get more action going."

Ozzie had his own announcement to make. "Our night out," he said. "I'm taking Freya to hear Ilse and my dad play." Their musical parents had already left for the evening to be the entertainment at a cafe.

"About time for you to start dating, after chasing each other around all this time! Isn't this your first night out, friends?" Norm goaded. His smudgy eyeglasses rattled when he put them down on the shuddering kitchen table.

Freya rolled her eyes. But it was true, she thought. Unless sailing a boat through a sea-attack together could be called a night out. Or standing at the feet of the Sphinx together with the full moon shining on them. Or getting out of jail together, or fighting for days to keep a sickly greenhouse alive. She decided to correct him: "Norm, we've been dating for months. We just had lots of chaperones all the time."

"Tonight, none," Ozzie decreed. "Fun only, tonight."

Freya put on a fancy tunic and they left.

They went and listened to Ozzie's father and Ilse sing popular love songs and some American country music favorites, along with old Icelandic songs that made the cafe customers join in loudly with them. They sang the Singer's song every half hour or so. Doug, his father, was good—her mother, too. She loved her mother's singing.

After a while, though, they slipped out and went to a club—to be anonymous and devil-may-care, to order a glass of wine and dance, to enjoy the freedom to be somewhere without disguises again and not worry about it, like other people. She tried hard to forget everything about the volcanoes and make it a happy time.

**

Ozzie stopped, breathless. He looked at her eyes, flashing with brilliant sparks, and loved her so much he started to laugh. Grand Galactic seemed funny. Even Raker was funny. And Seth? He was hilarious. He tugged her off the dance floor toward a table.

She sat and laughed at him, and with him, then tugged him back onto the dance floor again. They danced a slow one, wrapped around each other like high-school kids, but happy to be just that idiotic and drunk on each other. Everything went away: all but Freya and the music and them together.

The song was coming to a sweet corny end and they were going to dance every last note of it. Then the dance floor lurched. A few dancers screamed. The quaking threw most of them to the floor.

He held onto Freya; they held each other up. They sobered quickly. They were among the few who didn't fall over; weeks on the boat had given them sea-legs. But it was hard to laugh about it.

In thirty seconds an announcement took over the sound system: "By order of Reykjavik Fire and Police Council, all citizens who are not first responders please return to your homes immediately and take shelter..."

**

They walked home quickly on nightmarish shaking sidewalks. On their way a newsfeed rang in for both of them. Neither one wanted to, but both of them looked:

The World Council has ordered troops into Iceland and Malta to install managing governments, despite protests by the two countries. Informed sources say the move is

necessary "because both have shown that they are unable to protect their citizens from the threat of recent volcanoes."

CHAPTER THIRTY-SEVEN

Next morning the shaking had subsided but Ozzie and Freya were up early, unable to sleep. The news said Iceland was talking back loudly about the idea of someone arriving to take over running the country. The government disagreed with the World Council's promises of improved conditions, calling them "attractive bribes."

Freya brooded: "If friends of Raker's took over Iceland, they might be friendly to the idea of extraditing you and me, too. We have no boat to take off in, now... But Icelanders love independence too much. I don't know if it will even happen."

"Well, I'm sure you're right," Ozzie said, "—if people take over here, people who are Raker's buddies, there'll be less freedom."

"And we couldn't do what we're trying to do, then. We'd better work fast, huh?"

**

Ozzie found that Malo had sent a text sometime in the night:

> **Our families have spent some time remembering our very old hero-songs, and we have practiced them. We invited the war veterans to come and sing them with us, and also sing other hero songs they know. They have been here each night since then to sing.**
>
> **I have told Ahanith as we sang but maybe she didn't hear: tell her they like to sing her song. For them we call it "Brother Heroes."**

**

Ozzie's latest text from Diana only said,

> **I'll have to look at that ring. Will bring test equipment.**

Ilse and Doug knew that Diana intended to set up her office and her research here in Reykjavik for a few weeks. There was no word about when she would be arriving. And then, abruptly, a car brought her to the house.

Ozzie barely recognized her, she looked so... urban. She wore the kind of clothes he'd seen on many young London women: under her usual moss-green wool cape, a silver-and-black striped tunic, black leggings and boots. The heavy-lashed makeup went with her hair, which was sprayed a sort of plum color and pulled up into a trendy-looking clip.

"I've decided that being here, where we have the most intense symptoms of the problem, I might be able to gather more information and do it faster," she said.

She seemed to be unconscious of the way she looked. She beamed at them as she unpacked. So few clothes; *she travels like Freya*, he thought. When she caught a glimpse of herself in the mirror, though, she pointed dismissively toward her hair. "Disguise," she told them. "Washes out."

Ozzie was glad that Dad and Ilse were rehearsing with the band now, so they had time to get more details from Diana before they were interrupted. While Norm's radio station and Alexis and Norm's promotional machine continued to crank down in the basement, Ozzie and Freya sat on the spare-room bed and exchanged info with her as fast as they could.

Diana put the last of her equipment on the small writing desk against the bedroom wall, then landed lightly in the chair beside it. "What about the ring, Ozzie?"

He pulled the protective scrap of paper out of one cargo pocket and opened it to show her: inside the buttery gold of the ring there was the exposed dark core.

She leaned and looked. Long pause. "Not in there accidentally." The deliberate understatement made Ozzie grin.

He held it out to her to take, but the center was hot as his fingertips touched it. He stopped with the damaged ring in midair, and tested it with another fingertip.

She touched it too, eagerly, then cupped her hand to receive it.

On impulse, she moved it above her desk, where a little handful of metal paper fasteners lay scattered. They leaped upward instantly, two feet upward in a snap, and clung to the exposed black core of the ring.

"Whatever it is..." she began, and a slow smile began too.

"...it's something important." Ozzie nodded. It had to be, it had been so carefully concealed.

**

In the middle of the night sometime, Alexis shook Freya awake.

"Sorry. Couldn't sleep, Freya. I remembered the name of Seth's mother. The one he said is always drugged up like a zombie. And his sister, who died."

"Whuh?" Freya pulled the curls out of her eyes.

"The names came back to me, from when I worked with Seth, you know, on robotics: Linda, Ronna. For no good reason I remembered them and it woke me up. Ronna's the one who died from abuse by Seth—remember when I found his records on Mars?

"But then I couldn't sleep any more. I think that cat"—she waved her arm toward her sleeping bag on the floor, where the tortoiseshell cat sat licking one paw innocently—"kept on tickling my face. So I did a couple of number puzzles, and I still couldn't, and then I got this crazy idea of doing word-games with the two names and looking them up on the worldweb…"

"OK, but what…"

"Freya. There are dozens of companies that are owned by an 'international holding company,' whatever *that* is exactly, called Ronalin, Inc. Just look: if you search on it, see all the companies worldwide with that name? See, there's this whole group in the U.S. *including* that one in Oklahoma that is "testing" with the bad vibes! So all of the bad-vibe testing companies *could* be owned by Raker. And then if you look over on this—"

Freya didn't need to hear any more.

"Alexis, I see it all! Just strange enough to be the exact truth. Let's go get this to Diana for the investigator.—Yeah, I mean now!"

**

Diana stood behind her while Alexis announced to those assembled in the kitchen: "Ladees and gentlemen: Through a company called Ronalin, which pretends to be owned by two women, Raker is the sole owner of the little Oklahoma 'testing' company that's putting the bad vibes out," Alexis said. "And probably the owner of all 11 of the other ones like it, worldwide. We'll know the rest soon…"

"Could have guessed that," Norm yawned enormously. Seven a.m., pitch-dark here, and Ozzie was still struggling to wake up.

"Well, now it's *proved*," Alexis said, with her nose in the air. "That mega-sneak Raker! May the gods of the cosmos spit in his face. Using a dead daughter and a drugged wife as his covers. Let's get it out on the airwaves, Norm!"

Now he was waking up. "Yeah," he grinned. "Let's." They headed for the basement stairs. "Someone want to bring us down some coffee?"

**

As all seven of them rushed through breakfast a little later, Norm said: "Have you noticed how some people will always take something like the formula engraved in that ring and use it for something nasty? Like a. start a war, b. take over power, c. harass people?"

Doug nodded. "Always seems to happen. Someone has to make trouble."

"Sometimes I feel like I can imagine the future far out ahead of us," Diana said. "Maybe someday we'll contact another race that is much better than humans are. But as far as I can reach with my imagination I still don't see a time when sentient beings become perfect. So we may have to keep fighting to keep things free from abuse. Maybe always."

"Look at this, Ozzie! Everyone!" Alexis leaped up

from her seat at the table, holding her phone toward them. "Newsfeed: Seth caused big damage to the GG ship *Liberty*. And he got the axe!"

"Heyyyy, Sethie does his usual!" Norm grinned. "Sooo nice to see him *finally* get what he *usually* deserves."

Alexis read it out: "Captain's Apprentice Seth Raker, son of California Representative B. Arnold Raker, has been discharged by Grand Galactic... in response to Space City outcry, for 'negligence leading to loss of life.' Grand Galactic's spaceship *Liberty* successfully emergency-landed at Moon Colony Three after Apprentice Raker's failure to secure air-exchange equipment for takeoff caused the deaths of two hull workers..."

"Should be jailed for that," Norm said. "Probably won't." With a somber look, he added, "Ah alluz knew that boy'd come ta no good."

Must be another line from one of Norm's favorite flat-films. Freya watched Ozzie, waiting for him to speak. Doug and Ilse turned toward him too. He had the most right to feel triumphant.

Ozzie took it all in with a long head-shake. "Too bad about those guys who died. So someone had to *die* to wake everyone up? He almost killed Freya too, on Mars! Maybe now people will know enough not to trust him."

His eyes went back to the holovid he had playing on his phone below the table: recent launch footage from Moon Colony One.

No anger, no fierce victory cry. *You've changed, Ozzie,* Freya thought.

**

With just two days remaining before the Reykjavik Festival was to open, Ilse and Doug had gone to sing with the band downtown to promote the event. Diana appeared at the kitchen table, where Ozzie, Freya, Alexis and Norm were clustered, working and chatting.

"It's official, so I want to tell you all first: PII has just opened a new space-trading line," Diana said.

Ozzie gaped.

Norm looked up from a pile of data slivers in surprise. "Trading? I thought you PII guys were against getting rich."

His mother smiled at Norm. She seemed to actually be amused.

"No, there's nothing wrong with getting rich. But there's a lot wrong with getting rich by destroying life for others. Our trading line will be set up to trade in goods that make life better. Really better. For example, did you know that a certain isotope found on some asteroids seems to reverse all cancers? Strange but true."

"I heard something about that on a newsfeed," Alexis said. "Just last week."

"Right. Discovered a couple of years ago. Today it's being controlled by two huge Earth corporations who agree with each other to make only tiny amounts of it available at high prices. At those prices, only a small percentage of the people who need the isotope can afford it. So PII has quietly acquired exclusive trading rights to an entire *asteroid* of the substance. Billions of tons of it.

"And we've set up a company at Moon Colony Two to put it on the market and make it as cheap as aspirins or ice cream so anyone can have it. We'll make money, of course, because after a few years it will sell everywhere. And the money will pay for more research."

There it was again, Ozzie thought. More surprising stuff from her. Norm looked like some kind of light was turning on behind his eyes.

Alexis said, "You are going to be *fought* on that..." She must be thinking of Raker.

"We always are," Diana said. She gave Alexis an ironic smile. "We've been a thorn in the foot of these medieval companies from the start, so we're used to it. Same with the people who make money from them, like Raker and his friends."

"But then who will fund your new trading line, with so much opposition?" Alexis said.

"The gypsies."

"The gypsies!!??" Ozzie stood. He goggled at his mother's smug look, dumbstruck. How could gypsies fund a spaceline? In his mind he saw their tents, their barefoot children, their simple lives. Malo had probably broken his piggy bank to pull out those old rubies for electronic equipment.

"Malo has a lot of friends, Ozzie. They want *real* space travel to happen. And they have more to invest than you might guess. They and their ancestors have been saving up for this, waiting for this time to come. For at least four thousand years."

CHAPTER THIRTY-EIGHT

"OK, LET'S WARM THIS PLACE UP! Let's sing the Martian Growing Song again!"

Ozzie and Diana stood looking out over the plateaus and valleys of Hengill. Ozzie smelled sulfur and rock and the scent of something green. He turned back to see Freya and Alexis catching up slowly, taking the long route to look at hot pots and springs. An enormous stage was perched halfway down the opposite slope, facing the grassiest hillsides, with gigantic speakers, each as big as Freya's basement, that had been shipped in by a contributor for a sound system that could be heard all across the hillside and down the valley.

Norm had stayed here all night, sleeping between shifts as emcee. Now he sang a round of Ahanith's song to the audience himself, using the words that Ilse and Doug had made up to go with it. Then he said, "OK, if you don't want to keep listening to me, *you* sing it now! Louder!"

Halfway through the next round, he spied relief and shouted, "They're here! Come on out, Ilse and Doug!"

His father and Ilse loped onto the broad stage wearing cowboy hats and furry vests, waving. Ilse wore her delighted smile. Everything is perfect, couldn't be better, the smile said. The crowd roared.

His father and Ilse were astonishing. So show-biz. And the valley floor below them was already carpeted with people, 5000 of them. Most of the valley was carpeted with snow, too, so the people seemed to stay warm by sitting on blankets and chairs, wearing thick jackets, hats, and scarves. When they sang the space of the valley chimed and echoed. "Now that's a lot of people singing," Ozzie said.

Diana had parked her rented vehicle in a dirt lot marked with a hand-lettered sign that said "Sciences and Media." Ozzie began helping her set up the equipment for recording this contest: the music versus the quakes. So far, whether the crowd sang or not, everything quaked, steadily. Sometimes it shook even harder.

Like now: the valley floor shuddered a couple of times. There was a wave of screams. "No, no," Norm called over the screams. "Save that for singing—Let's go, folks!" Ilse and Doug hit a few hard chords and upped the volume, taking the tide of voices with them.

Ozzie reached for the ring at his throat, then remembered: it was out of his hands now. For testing. He wondered briefly if having it gone was making the tremors worse here. If the ring held the formula, might it not also hold the solution? Or clues to it? But the news this morning said the quaking was getting worse everywhere.

People arrived steadily, trickling down into the valley: clambering and sliding downhill over rocks, snow and grass, carrying backpacks and food-storage chests, some of them even holding the tethers of expensive hover-coolers that bobbed along above the ground behind them.

Steam rose in little clouds from the fissures and hot springs that speckled this slope, as well as the ones across the wide valley. Around the springs the grass was green. Ozzie could hear the hiss and bubble of the nearest one. How near to the surface would the lava have to be, to be boiling water on a winter day? He was afraid he could guess.

"It's steaming from all those little springs, "Ozzie said to Freya when she returned. "And there was no steam before?"

"Not exactly," she said. "It's been steaming forever. At least forever as far as I know. And some of them are just steam with no spring. Many places in Iceland steam, you know? But about two years ago this place began to belch *lots* of smoke and steam, not just the usual. So close to Reykjavik, it was scary. Every few months since then, it has done a big fit of belching—and no worse than that, but still scary."

IN THE RING

**

A popular Icelandic group left the stage. The whistling and applause died down.

Norm's voice filled the valley again: "Ladees and gentlemen, I want to welcome you all again, and all of you who are just arriving, to Reykjavik Rocks, presented by Radio Reykjavik—the music festival where you are one of the stars. We need you to sing, all of you!

"As you know, we have some music brought from Mars that has a track record for correcting disharmonious vibrations. Yes, I mean it. That sounds way-out, huh? But this music has made dead plants grow and dry rivers flow wherever it has been played. We are here to play it and sing it, and all the other music we want, together!

"This festival happened when Ilse Arnsdottir and Doug Reed decided to get more singers to join them in singing the mysterious Martian music that was brought from the Red Planet by scientists"—he waved wildly in the direction of Ozzie and the others—"who happen to be here today!"

Alexis looked horrified. The crowd cheered.

"When Ilse and Doug's holo-posts went viral, the result was what you see happening here...

"Are you ready for another set by these two, Ilse and Doug, Iceland's newest voices?"

The crowd bellowed encouragingly.

Ilse and Doug Reed strode onto the stage, waving.

**

After Doug and Ilse's first two songs for the set, when Alexis and Freya went downhill to look at another hot spring, Ozzie said to his mother:

"I'm still amazed that you and Dad are friends."

"I know, Ozzie. Life is complicated sometimes, isn't it. Love is too. I want you to know: I *loved* him. That's how we were lucky and got you."

He felt a twinge of sorrow for all he had missed, but it dissolved pretty fast. Nice to have her around now. "And it's actually better that you're apart?"

"Do you really wonder? Look how happy he is."

Ozzie gazed across the distance at Ilse and his father, leaning toward each other as they sang, moving together to the rhythm of the music and lost in their song, as if there was no one else needed at all in their world.

"No. You're right, nothing to wonder about." *Except you, my mother.*

**

The sun was setting already on this short-lived winter day. Every minute, hundreds of arriving Festival guests streamed over the brow of Hengill and poured down into the valley.

"Welcome to Reykjavik Rocks," Norm intoned again from his microphone at the side of the giant stage. The next band was setting up. "You've just heard Ilse and Doug, whose great idea it was to have this festival. Our excellent media people, who know so much, say that there are at least 40,000 of you here now. And dozens of top-flight bands. Let's give a welcome to all of our friends who have arrived during this set!"

The crowd roared.

"When the holo-posts about the Growing Song went viral, more happened than just this little hometown festival." He chortled, as the audience roared again. "Those posts have made the noise happen worldwide: by worldweb, by holo-channel, in stadiums and rock venues. All over Earth people are arriving right now to pour out the music. In New York and Paris and Honolulu they have been singing since last night. By this morning 65 separate venues were singing live, and you and I are being broadcast, right now, at over 50,000 public listening sites…"

They roared again.

"Today hundreds of world-class musicians are here to sing with you. And I mean WITH you, folks. Top musicians worldwide have staked their lives to be here at the center of the danger and sing. So have you—and give yourselves a yell for that…"

They yelled, a long time.

"Hey. Hey. Don't wear your voices out. We've gotta help each other sing. Right now the volcanoes are trying to unzip Iceland from North to South, from the Arctic to the North Sea, and we'll be here singing all day and all night long until they stop!"

It went on hour after hour. The excitement was beyond electric. Ozzie had never seen so many people in one place. Helicopters hovered, big commercial planes went back and forth overhead to and from the airport, while cars and copters and many private planes brought the arriving musicians and their swelling audience to the Festival.

One weird vehicle showed up near the rim of the valley, with spotlights on it. "Hey, look," Alexis pointed. "It's our OverSeas delivery guy!" She texted Norm.

"Hey you all," Norm gave a broad wave to his audience. "Just got a message in. See that winged truck up above there? That's a friend of Reykjavik, paying me back for a favor by flying in some of your favorite stars for this show. Thank you, Thoren Bisk! Thank him and these brave musicians!

"—OK, folks, I think they heard you yell—Now show them what you can do with the song!"

The music began again. They roared the Growing Song.

The sound system had been doubled since this morning. Ozzie wondered where the extra speakers had come from. Right now, the whole broad valley was full of bodies, all of them singing.

**

Without warning, the earth shook hard. Ozzie held his mother's arm. She wasn't the kind of person who fell over easily, but he thought he'd better. A chorus of screams swept across the valley.

"Cheer up, everyone," Norm's voice filled the air again. "This is more fun than taxes or cleaning house."

The next band was on the stage now, starting up. "This is scary," Alexis said. She looked longingly out across the valley at the stage and Norm.

"Yeah," Freya nodded. Ozzie knew what she was thinking: *Hengill. Lying in wait for us.*

"I need to be here to help Norm. Every day. He might not make it if I don't," Alexis joked. No one laughed. She pursed her lips.

"Tomorrow," Diana said, "I'll start arriving here early, daily, to take sound readings and info from Norm and relay them to the PII studio. You can ride with me, Alexis."

She left them to go set up her instruments and get her final measurements for the day.

**

"Newsfeed, Raker," Alexis said to Ozzie and Freya with disgust. They sat in the lee of a large boulder to get out of the wind. Someone had handed them coffee, and the hot stuff warmed her hands.

Freya read from the phone aloud:

World Council troops will be landing in Reykjavik to shut down Reykjavik Music Festival broadcast of this music, despite protests by the Icelandic government. The Council claims that the music is causing brain damage in New Zealand infants—

"He is so creative," Alexis said glumly. The ground shook a little extra right then.

—Scientific briefs have been sent to all governments worldwide urging them to get the music off radio stations and out of public places...

Ozzie said, "It doesn't seem to matter that the New Zealand Department of Sciences has proved the brain damage claim is false."

"But here's more," Alexis said. "Look ..." She and Freya huddled over her holoscreen.

Oklahoma City and San Francisco have sent delegations to the Council to protest Raker's claim. There are also marches today in protest at the Sorbonne in Paris, at Oxbridge, Berkford and Harvard, Columbia, New York University, Swarthmore, at 2000 U.S. and 1000 European colleges and universities...

And the Veteran Voters of America have hit the streets to demand that the music be allowed. So have...

"Holy smoke," Ozzie said reverently. He was following the feed on his phone too, and he tossed a holovid out in front of them. There before them, armed soldiers filed into Reykjavik harbor. The holo showed them on guard at all ports of entry: Reykjavik airport, the entrance to Seydisfjordur harbor...

"But the uniforms are Iceland's!" Freya was confused.

The holovid voice said, "... At all Icelandic entry points local citizen defense groups, a legacy from the Wars, have moved into position to guard against entry by World Council troops."

Freya sighed, closed her eyes and smiled. Raker had helped them out again! *Now we're getting the kind of allies we need.* For example, the Icelanders themselves.

**

Diana shifted the strap of the instrument case off her shoulder and sat beside them, there where the huge boulder blocked the wind. "All OK?" she asked.

She smiled at their news.

Hers was next: Some real improvement in the worldwide seismic graphs, showing up since the largest festivals began last night. And a new PII scientific investigator had just arrived here in Iceland to work with her.

"Just landed, he texts me. But he says there was a fire at PII. Damaged the recording equipment. What person outside the company could have known about the recording studio? Our people are loyal."

"How 'bout the private investigator who turned out to be a traitor?"

"Maybe."

"Can they fix it?"

"Of course. They are working fast to restore it with emergency backup equipment. Just costs us more time and money."

**

Diana's rugged vehicle stood in front of Freya's house with its doors open. Freya helped him unload the last satchels of equipment to carry them into the house for the night. Tiny grains of falling snow melted on their faces, leaving little jewel-like drops on Freya's nose. She said, "Game standings, Ozzie: Hate to say it, but there's a newsfeed from the twisted world of Raker again. A list of countries taken over: Malta, Costa Rica, Saint Lucia, Seychelles, Marshall Islands, Cape Verde…"

Ozzie slammed the metal doors shut. He raked his fingers through his hair, throwing off snow and water. It was painful how much power the guy had. The people on their side seemed powerless by comparison. "Damn, Freya. What's it going to take to stop that maniac?"

CHAPTER THIRTY-NINE

THE NEW PII RESEARCHER who had come to Reykjavik was dressed unobtrusively. He stayed at a local hotel. He was introduced as Dr. Nimr, which sounded to Ozzie like a gypsy surname—not that he'd heard millions of them. He had dark eyes and hair, and flashing teeth. His looks reminded Ozzie of something.

Later, it came to him so he asked her.

"Yes," she said. "You're right, he's related to Malo. In fact, Dr. Nimr is Malo's son. One of many," she added. Ozzie was startled. A son of Malo, who was so old? This guy seemed kind of young for that. But then, so did Malo.

**

That night, Diana was running electrical tests on the ring when Ozzie stopped in to watch for a minute. He stood in the doorway of the little bedroom, not wanting to disturb her work.

Dr. Nimr, whose desk was a lapboard and a chair

under the window, looked up to gaze quietly at her sitting there hunched over her test. "You may be cold," he said.

She was too absorbed to answer. He rose and set his board on the bed quilt among scattered papers and instruments. He took the cape from the back of her chair and draped it over her shoulders, moving her hair to lie on top of the collar.

She shrugged gratefully into the warmth as he did so, and then woke from her concentration and turned with her eyes lit, to look up at him. His own eyes caught fire from hers and he smoothed her hair briefly, then turned away.

Did they say anything? There were no words that he could hear. *They sure spoke plenty, though.* Ozzie had never heard so much said by eyes. She and this guy must like each other.

Then it dawned on him.

**

His mother was married to a gypsy. He couldn't think of anything more perfect.

"Yes, we're married, Ozzie. Different names, rarely travel together, for safety. If one of us is lost the other can keep PII going. Now that you know, you must help us keep this secret."

He nodded. Another secret. Of course he could keep it; he had practiced for years.

IN THE RING

**

Freya looked out over the valley. There were so many Hengill visitors, now, that bands were set up on small extra stages all over the area to play while the crowd rested between long rounds of the Growing Song. Up here on the rim the sounds of the various bands made a clashing local background noise that was only overpowered by the huge sound from the main stage down below.

When it was almost time again, the supporting bands would get the seated thousands warmed up and ready to give it all they had on the Growing Song. And when Ahanith's song wasn't led by Ilse and Doug, a band leader on one of the other stages did the job, or joined Norm on the main voice line and they led the singing together.

Alexis was down there with Norm all day now, whenever he was on stage. And with him whenever he wasn't, too, Freya thought. *Taking care of him.*

Norm kept himself amused, and kept his word to Freya, by giving the Hengill audience, and his Radio Reykjavik listeners, all the news about Raker and the Council. He announced each new takeover of a small nation, each new attempt to close the Festival. That was today's news topic. "...so let's send Raker a message, people, in case he's listening, about how *impossible* it would be to stop Reyjavik Rocks."

The valleyful of people roared.

When Freya asked, a Festival volunteer told her that most of them managed to keep on roaring by sleeping part of each 24 hours in thermal campgear, or in cars in the parking lots, and eating whatever food they had brought or could send a friend out to pick up. Reykjavik residents, stores and eateries were doing a booming business in food, cooked, sold, and even delivered, round the clock, to feed the thousands of volunteer singers.

**

Diana had just recorded more measurements. Bits of rock rattled downhill ahead of her as she came from the little Science and Media platform, at the rim of the valley, to join Ozzie and Freya on the protected side of their windbreaking rock.

She answered their unspoken questions with a headshake: "We're gaining, just not gaining on it fast enough." She blew on her hands to warm them, then put her gloves back on.

At the Las Cruces Campground, Malo had begun to use his phone to send Freya or Ozzie encouraging updates from the gypsy network. How he got the news as fast as he did was hard for either of them to understand.

Right now Freya picked up his latest: Malo reported that in the U.S. campgrounds, the gypsies had started getting the rest of the residents to sing with them, the ones who weren't veterans or students. "We call them the Lost People," he said. "They have no one and nothing much. We ask them to help us. Sometimes we just gather them up and feed them and tell them to start singing, but then they get interested and come by themselves after a while."

**

She looked out over the steaming valley. Snow was melting. Was it because of the heat of thousands of bodies? Or the heat of melted rock, churning just underground?

She thought again of the PII plan to start a space-trading line.

Norm and Alexis planned to finish school.

She and Ozzie planned to have happy lives, adventuring.

Malo planned...whatever it was that Malo planned.

And no one planned to be blown up by volcanoes sometime soon. Probably the Atlanteans, before the flood wiped out Atlantis, were busy planning things too.

The thought terrified her.

CHAPTER FORTY

BY 8 P.M. ON FESTIVAL DAY 7 all of them had made it back to the house for something hot to eat. It was Ozzie's turn to cook—his idea because Norm and Alexis had been exposed to the punishing wind all day long. Norm needed help when he cooked, anyway, to make anything but pizza and enchiladas.

At the Festival site an hour ago, Norm had put the Singer's voice, doing her song, on automatic with some overnight programming. Now he clomped cheerfully into the kitchen with Ilse and Doug, shaking off snow, ready for supper.

Freya helped Ozzie ladle vegetables into bowls and lift fish onto plates. Was it Freya's imagination?—that the shaking was even rougher now than this morning when she was the cook? The little herb-pots on the window sill rattled nervously.

She shook her head, then looked at Ozzie. He was watching her.

They sat anywhere to eat; not enough chairs at the table for eight. Ilse and Doug turned in their plates with thanks to the chef and went directly to bed. "Another long day in the cold air tomorrow!" Doug said. But they were pink-cheeked and happy about it.

Freya was grateful for the warmth in here, and the hot food. And even, surprisingly, for the potful of tea. Norm and Ozzie prowled looking for seconds.

As soon as Ilse and Doug disappeared, Diana looked at Dr. Nimr, then addressed the rest of them urgently:

"We have something to tell you all. From the private investigator at PII, we now have the full story on what Raker has been trying to do! We don't have a minute to lose. Doug and Ilse don't need to hear this now—need rest more—but you four should know."

Norm and Ozzie quickly brought their plates back to the table. Ahanith's eyes appeared in the doorway to the hall.

"Our new private investigator got hold of papers and files from a former employee of Representative Raker's. And some papers that Raker has filed that connect him to certain causes as a backer and an investor. The list of them might interest you sometime. But one of them is the vibrational "testing" in Oklahoma, which is also being done in eleven other locations around the world, by eleven other companies. Raker owns all twelve companies—that's right, Alexis.

"I know *that's* not a big surprise to any of you. But

Raker turns out to be involved in covering up the growing volcano problem, disguising the cause of it, and putting out false information, for a hidden reason. Now we know what the whole plan is—his own records say it. But his records also show that it's not just Raker's plan; he's just the tool of some backers who must be pretty scary."

"Thought he must be *somebody's* tool!" Norm chortled.

"They plan to do it by shaking up Earth, literally, to control populations by extortion—"

"Extortion? Threatening people with harm unless they pay up..." Alexis frowned.

"Yes. Their extortion strategy is to get Earth people so frightened they accept rulership by a government that promises them safety."

"Our private investigator found a document that included something worse: they plan to make disruption and damage so extensive that no one on Earth can trade with the Moon. That way they can force themselves into control of the lunar colonies, then Mars. They plan to wreck Earth enough to make people here beg for it to stop, then they can run Earth and Mars from the Moon—and prevent the Moon and Mars colonies from freely settling and trading.

"All this is to keep people 'under control' so they alone can run things, they alone can profit and be free."

Ahanith spat, and the noise was so loud it made

Alexis jump. [Destroyers!] Freya heard her hiss. Her eyes flared, then completely disappeared.

"Now that we have proof," Diana said, "we can begin to stop them, tonight."

Dr. Nimr nodded at her. They rose. To Freya, Diana and Dr. Nimr towered like a god and goddess, there was so much intensity in their eyes.

**

Ozzie rose too, and took plates from them. He was horrified. "What they're doing would end 'free trade and free travel.'" He struggled even to say it. It was every spacer's slogan—and what the Moon Colony Rebellion had been all about.

He tried to line it all up fast in his mind: "My grandfather's formula was found in his files at Grand Galactic years ago. It was probably figured out by GG scientists quietly: they tested it, decided it wouldn't be allowed for outright destruction, but it could be used for sneaky destruction, since it would seem to come from no source... So the extortionists must have gotten word of GG's technology. But who are they? Who would pay Raker and a space-trading company to do this?"

Freya stopped stacking the rest of the plates. Her eyes narrowed. "Yeah. Who are they exactly?" she asked at Diana's disappearing back.

Diana turned. "We don't know; there have been attempts like this earlier, before and during the Wars, and it's not clear who is behind them. It would be enough, for now, just to stop them one more time." She and Dr. Nimr disappeared. Their office door closed quickly.

Ozzie sagged into his chair again.

Norm drifted toward the basement stairs, sleepily. "Fabulous investigation, PII," he called after them. "Go gettem! Confusion to Raker!" Alexis shushed him and herded him forward down the steps.

**

Ozzie and Freya talked softly while they cleaned up. She was glad for the time to talk. She could hear Norm and Alexis below, murmuring excitedly and sorting out the mess in the basement room again so they all could find places to sleep.

Ozzie said: "So Diana and Dr. Nimr will blow the whistle on Raker's plot... Sure, that will help, Freya, as long as it means somebody listens to them and stops him! Someone needs to stop him. But I just don't see what the ring has to do with this. It must have something to do with this. Why the ring?"

"Dr. Tersey saved the formula by engraving it in the ring..."

"But he knew someone there at GG had already been through his research files and probably got hold of the formula.—So why would he save the formula, if the cat was already out of the bag?" From the floor he caught the gaze of the tortoiseshell cat. [Scuse the expression—]

[No offense taken.]

"I don't know. Maybe as a warning to us—which it is, isn't it?"

Ozzie looked dissatisfied. "I wish the ring was an antidote for the shaking."

"Is that what you're thinking about? Well, I wish too. Maybe it will be! For now, we'd better keep doing what we're doing." He looked blank, so she added: "The music."

He sighed. He still looked dissatisfied.

**

Responding to the flick of her eyes, Ozzie shut Diana's office door behind him.

Pale moonlight sifted through a little window above her desk. He knew their work to expose Raker's crimes had been finished when Dr. Nimr left two hours ago. Now she must be waiting for the result. Like him, it seemed, she couldn't sleep. Sleeplessness had touched Norm and Alexis too, for a few hours, and then their noises wakened him; but they were silent now.

Diana had finished testing the ring. "The instruments we have locally are only enough to tell me that it's a core of carbon in nanotube form, formed or bundled into a 'rope' of carbon so strong it can hardly be cut. Makes steel seem like putty by comparison.

"But this stuff is so carefully embedded in such a

strange place, in a gold ring—and a ring with a formula on the outside, too!—that there must be more your grandfather was trying to say.

"A special courier is going to take the ring safely to our New Mexico lab for more discovery," she said. "It must be done quickly; what if this is the way to undo the vibrations?"

He nodded.

"It's too precious to be taken by anyone less trustworthy, so—" Her voice became a miniscule whisper. "—Dr. Nimr will take the ring back to our labs."

Ozzie stared at her. "The risk will be huge, won't it? Considering what Raker knows that PII knows…"

"One of our planes will pick him up tonight, on its routine supply run to our Reykjavik observation station here. Dressed as a mechanic who is shuttling back to SpacePort USA. He and the ring should be safe."

The whole house was shaking. Hengill must be pumping lava beneath the streets outside, by now. There was so little likelihood that any of them, any of their little group who stayed, would survive.

How could he say this so she understood? "Diana. Mom. You should go back to New Mexico with him, with Dr. Nimr. Survive with him, make a better world and all." He saw the refusal in her face, so he urged: "Now, when you're happy, you should stay together."

She looked at him and smiled. He was surprised to see tears swell in her eyes, especially since they seemed to be for him.

"Ozzie, we need to make this happen here. He would not be leaving except for the importance of the mission to go and test the ring. And one of us should be here. If we don't succeed, humankind on Earth has poor odds of survival anyway, even in New Mexico."

Ozzie thought of Freya's tale: about how Ahanith chose to go on living instead of dying with her Martian husband, her Other. Maybe Diana had the same instinct: one at least might survive to continue the work.

"We need to stay and make this happen somehow, that's all," she said. "We have to succeed."

**

Ozzie put on the kettle for some tea or something. The refrigerator vibrated, rocking a little from side to side. With Norm and Alexis and Diana needed at Hengill, he and Freya had volunteered to stay at the house today to receive all the contest entries, collect the music, save the entry records, and send the music in volleys to Norm at the Festival site.

There Norm would play the winning entries from Venezuela, Israel, Canada, Korea, and so on while the crowd rested their own voices. Norm had figured out a fast, easy way to get the power of the contest entries measured against each other: get PII to see what kind of difference each one made at the Icelandic PII test station.

Ozzie and Freya had just sent off 45 contest entries.

There would probably be more still, this afternoon.

He found her sitting in the little living room sort of place near the kitchen. That golden-eyed cat purred comfortably on her lap. The restfulness that surrounded Freya made him want to sit in it, beside her. So he did.

"What about Ahanith, Freya? If the people on Earth are mostly killed and the place becomes a trash-heap for a while, won't it be good to have her survive to sing it back to life eventually? Maybe we should send her back with Dr. Nimr so she's safe."

"Ozzie, she can't die anyway. I mean, she has no body that can die."

"Right... Well, then maybe she would be traumatized here and go to sleep again for who knows how long."

While she considered this, Freya's eyes flashed and cooled, flashed and cooled, softly like embers. "Guess we should ask her, huh? It's really up to her anyway."

They went to find the Singer.

She was nowhere near the goatskin flask. "Maybe she's already gone," Ozzie muttered.

A cool wind passed his cheek, moving fast. Ahanith coalesced before them, more solid-looking than they had ever seen: her head crowned by a sort of decorative headpiece like queen Nefertiti of Egypt, wearing a long white tunic and leggings, her eyes vivid turquoise green and her skin pale-green: the color of the back of a new leaf. Ozzie's jaw dropped. She was almost too beautiful to look at.

[Malo has sent word,] she said. [A poem.]

"For you!" Freya breathed, out loud. Ozzie knew what she was thinking: This was epic, a poem from her love.

[Yes, for me. For you both, too, I think. He says:

> Even if I lose you again
> I will always find you
> Even if I lose you
> for a while
> I will always find you
> in a while
> I will always find you
> again.]

[And did you answer?] Freya's thought seemed awed to him, like a whisper.

[I sent him a song. From the Festival Norm helped me send the sound and the words. The words say,

> Now that I have found you
> I will never lose you.
> I will never lose you,
> There's no need to find me
> You can never lose me
> again.]

**

Ozzie thought of Ahanith as they all ate supper. About the way she looked a little while ago when she recited the words of the song she'd made for Malo. Ozzie was used to thinking of the Singer as little, but today, even when her eyes were about the height of Freya's, she seemed to be as tall as the Statue of Liberty, and her eyes were as deep as you could look between stars into space.

Their table was less full now than at mealtimes a few days ago. Dr. Nimr was gone. And as of today Doug and Ilse had moved to stay with the other Festival musicians in makeshift lodges that had been thrown together near the Hengill site.

"I'll clear the table when we're finished," Alexis said. "And Norm is dying to do the dishes."

Then the ground shuddered heavily. Ozzie stood up quickly, at attention. He heard a roar coming from somewhere.

The house began to shake outright, hard.

CHAPTER FORTY-ONE

ANYTHING THAT WAS LOOSE IN THE HOUSE clanged or rattled or fell to the floor. *Will my grandmother's house shake down to the ground?* Freya thought desperately. It made no sense to worry about such a thing, but the idea of having no house anywhere suddenly bewildered her. Then it shook again, harder still: the width of a fist one way, then the other.

They all held onto things and stared for a few seconds. Diana looked at Ozzie, then Freya. "Let's get out of here," she said. "Safer out in the open I think."

"I need to be at the Festival," Norm's eyes showed that he was already there.

Ozzie nodded. "We'd all better just stay there around the clock, starting now. We can do more to help, there." Alexis dumped the dishes in the sink, and she and Norm ran to pack, right behind Diana.

"Wait! Ozzie!" Freya showed Ozzie the title of the holo newsfeed that was just ringing in:

Pasadena Rep. Raker Confesses to Volcano Crimes Conspiracy

Then there was a holovid clip with a caption that said:

Oklahoma state troopers surrounding the fracking facility south of Oklahoma City and its secret sister facility, a site emitting the destructive vibrations that have been named as the cause of the worldwide storm of volcanic outbursts.

The places were closed, fenced, posted with legal signs, and guarded. There was a holo of chiefs of police of Oklahoma City, the Cherokee Nation, and the City of Tulsa receiving copies of the order to shut down and guard both sites. In the background of the holovid dozens of armed guards stood, looking dangerous.

"YESSSS!" Ozzie roared. "Norm! Listentothis!" Alexis and Norm's feet pounded back up the basement stairs. Freya put it on Speaker and maxed the volume.

A voice went on:

...local residents have blamed the fracking facility for the Oklahoma volcanic eruption and the tremors throughout the state. But it was not the fracking facility itself that caused the problem. Masked by the vibration of the fracking

was the vibration of this neighboring plant, where specialized electronic equipment was used to put a persistent discordant vibration down into the fractured layers of the earth, causing seismic disruptions, volcanoes, and magma flows.

And not just in Oklahoma. There are a total of twelve sites, located on all seven continents, where these destructive vibrations have been manufactured and pumped into the earth. All twelve have been located and reported by Premier Independent Investigators, known as PII, a U.S. independent research company that has earned a reputation for environmental investigation...

All 12 sites have now been shut down by order of their local governments, with local police or military organizations guarding each of them. New Zealand has provided Special Forces as first-line guards at the facility in Antarctica...

Freya felt like she had electricity in her veins. "Did you hear all that, Alexis?"

"Yeah," a slow smile started up one side of her face.

"What! Tell me."

"Don't tell anyone you don't have to; not yet. Norm and I leaked this story anonymously to our whole list of radio stations. In the middle of the night, last night."

"You leaked this?—Well, is it true? Did he confess or not? Is he arrested or not?"

"Um, he confessed after this. He is arrested now. But he wasn't then. Diana sent us a bunch of info during the night and we just went with it. Doing what you asked, Freya: cut off his power sources! My story said 'Raker Linked to ...' with a note that they should check before they ran it and see if he had confessed yet. I think some of them just saw 'confessed' and ran the story because it sounded juicy...

"After our recent news releases about him, and Norm's broadcasts, I guess he didn't sound like Mr. Perfect to them anymore.—I didn't lie, but it had a spin on it that made the cops hot to get him arrested—and by the time he was, the story was running with 'confessed' and they had pictures—and who knows which happened first?" She stopped breathlessly, shrugging, and giggled.

Freya stared at her, open-mouthed. Then she threw her head back and laughed, slapping the table. "Well, that's the most fair thing I've ever seen happen to this guy."

**

"Good news!" Ozzie pounded on Diana's door. She had to see this newsfeed.

"Hey Ozzie..." Norm's voice sounded pale. Ozzie looked behind him at Norm and saw that his face matched the voice.

"What?" he turned.

"Look: Pasadena. Just came in now. Still trying to reach my family." He fiddled with the phone ineffectively, trying to bring something up to show Ozzie.

"Just tell me!"

"Raker's house went into a fault that zigzagged through Pasadena. A big crack opened... Pasadena is cracked in half, buddy!"

He sounded like he was coming apart, himself.

"There was a map, think the crack missed my house, but I don't know, and the schools, and even Uncle Chang's—his is just a couple of blocks from Raker's. Raker's house fell in. I'm trying to call my family...Here it is, lemme read you:

> **At the neighborhood Dairy Swirl, a ten-foot-tall decorative ice cream cone, part of the structure, was left tottering at edge of the fault after it slammed shut again on the rest of the 100-year-old ice cream stand, which had fallen into the gap—**

Norm sounded a little delirious, reading off all this junk. "—The Dairy Swirl? Aww, bummer, Oz—

> **There at the Dairy Swirl, a high-school student closing down the store climbed up one side of the cone to get out of the crevice, just before it**

closed again to crush the rest of the building. She was found and interviewed later with her apron still on, sitting in shock at a hoverbus stop half a block away from the site...

... said the sticky ice cream on her hands helped her grip the slippery sides of the giant ice cream cone and cling to the curl through the last big shake.

—Ha! Ridiculous!"

Norm was trying to make himself laugh. But he sagged into panic again: "'Representative Raker, who was arrested and released on bail just yesterday, was not at home... Raker and his family have not been located...'"

"I get it, Norm. Holy crap—"

"—But my family, I don't know if they're—" He paused.

Then he took a huge breath and let it shudder out. "Text in: My mom! Scared but safe, she says. Whoooo. Just a minute, let me tell them to get out of there..."

He looked really bad. So Ozzie didn't say it: *Where are you gonna tell them to go, Norm?*

**

Ozzie knocked again harder. Diana answered.

"Yes, I got the news," she said. "Waited to hear all of it." The house shook again.

She might have been the only scientific-sounding one there. The others who gathered in the hallway were wildly emotional; Freya herself felt like leaping out the door and running. Crowded together, their energy was intense.

Diana looked at them all, then said: "That's right, the cause has been shut off and Raker's arrested!" She smiled grimly. "We've done well. We have won the fight to stop damaging the earth's crust. *The planet will right itself; life here will survive.*"

"Yeah, baby!" Norm howled. Alexis initiated a group hug, sobbing happily.

But Ozzie must have heard something in Diana's voice, because he stood back and waited, warily gazing at her.

"But listen. Listen, friends! The waves are still moving through the earth's crust. What has already been set in motion is still in motion. Until the motion stops."

Freya saw the look on her face. *Of course: the earth will keep heaving until it settles and the magma cools. The most fragile places are still fragile. The lava is still sloshing and spewing out of holes and cracks...*

The house shook again, demonstrating the lesson. Each of them braced against the sides of the hallway or each other.

Iceland had to be one of the worst places to be, right now. Not over yet.

"Isn't there any way to stop the motion, now that no

one's causing it?" Alexis asked. She was like a frightened girl, not the smart scientific Alexis. Freya saw her nightmare all around her: vivid tongues of crackling flame.

Diana spoke quickly and soberly, "You know that's what we've been trying to find. And you know what we have found so far: no instant antidote, or switch we can turn off—"

"But the ring might—" Ozzie said.

Diana shrugged. "Yes, it might. We just don't know yet."

"And here we are on a fragile place—" Alexis mourned.

"—And people all over the earth are on fragile places—" Norm stared at Diana.

"—Until it all stops," Ozzie said. "How long will that take?"

She shook her head slowly, as if she were trying to think. "We have to go, quickly..."

The floor heaved. Freya felt sick.

Diana held onto the doorframe. "Here's what today's Earth science would predict: The worst motion, weeks. More motion for a month. For it to return to normal, months, maybe years..."

Ozzie stared at nothing and said, "By then, who knows how wrecked the planet will be?"

Norm recited lamely: "It's Physics 101: A body in motion will tend to remain in motion, unless acted upon by an outside force. We just need an outside force..."

"Well, we *do* have an outside force," Freya said. "The music."

They all just looked at her. Hopelessly, except for Diana. She saw that Diana's eyes had a surprising glimmer of light in them.

Ahanith, as if she had been summoned, now stood queenly and tall beside Diana in their small circle, her eyes taking all of them in. [We should not give up,] she said.

Ozzie nodded slowly. He said, "Well, we all have to choose for ourselves. Anyone want to leave? There's still time to catch a boat or a plane..." He gazed at Alexis, then Norm. "And it's OK. You can go if you want. If you need to." He looked at his mother, and then looked away. Then at Freya. And his eyes went to the floor.

Freya got it. No emotional appeals. She wasn't doing it to him, so he wouldn't do it to her. They both knew what they were deciding.

"What, and miss my biggest DJ nights ever? No way," Norm said.

Alexis looked terrified. "I'll stay with Norm," she nodded. She was very pale.

Freya looked at Diana, then Ozzie. Two pairs of gray eyes looked back at her.

With you, I can do anything. She said, "OK. Let's do what we need to do, then."

But she knew Ozzie had already heard her.

CHAPTER FORTY-TWO

THEY DROVE IN DARKNESS toward Hengill. Their backpacks and Diana's equipment were stuffed into the back of her vehicle. He read the newsfeeds to Diana and Freya, Norm and Alexis as Diana drove, although Norm and Alexis were silent and probably asleep back there. They all needed it, actually.

Out here where man-made lighting didn't interfere, he saw a rippling ribbon of colored light above the line of hills. He pointed. "Northern lights!"

Diana nodded and smiled at him briefly. She looked worried.

He played one live newsfeed for them. The voice said, "All commercial airlines ceased flying a few hours ago when smoke began to billow from a crack at one side of the Hengill site. Many Festival visitors left the music sites and swarmed to catch the last planes or commercial boats. And after more than a week at Hengill many well-known bands have left the area to go to other engagements, but others are still arriving by sea to take their place.

"It seems that right now everyone with a death-wish is flooding into Rekjavik by private transportation somehow; the Festival site is full and worried Icelandic emergency crews are urging newcomers not to land.

"Some of the most suicidal have arrived in the night in little boats and planes that managed to be swamped by the rough seas and drown, or crash into lava headlands, thus accomplishing their missions more or less..."

Ozzie scowled and turned it off.

He surveyed the feeds again. "There are over two thousand festivals going, worldwide," he told Freya and Diana. "I see the holovids, hundreds of them listed here. Hopefully in safer places than this one."

He flipped the listings on his screen upward with his thumb till he got to the little amateur vids that were coming in: 51,378 had streamed in, just in the last 24 hours...

One showed something that made Ozzie's throat tighten. Someone had pasted together live holovid clips from America, ones posted by people in Los Angeles, New York, and somewhere in Massachusetts, Montana, Texas:

There was one of hundreds of people stopped at a roadside out somewhere in farm country, raising candles and matches that lit the falling snow, tuned to the few phones with power left among them, singing.

There was a holovid of a traffic jam in some huge city, with people standing outside their cars, raising phone-lights and singing. *(The camera is shaking. Is that because the ground is shaking there, too?)*

There were short holos from some Albuquerque and Portland and Seattle bars, where the barkeepers doled out the drinks sparingly while their patrons sang. The bars offered free coffee to keep the patrons awake. "Just keep singing!" one bartender ordered his customers.

If the little holos were intense, the world-scale holos were large-scale amazing:

"Listen!" He tossed the 3D image up into the air in the front seat so they could get most of it as they went: it was a holovid of a muezzin singer in Egypt who did a heartbreaking rendition of Ahanith's song and then crowd-dived from a second-floor balcony. "They're carrying him on their shoulders through the streets, and the street's full of people singing with him...can you hear?"

"And here's another one just in: 2000 people from a church sandbagging an overflowing river out in the western U.S. somewhere, singing the song..." He upped the volume so the three-part harmony filled the vehicle till it was over.

Freya's eyes were already electric but he didn't want to stop: "Here, another one: London, Freya, last night!" He tossed a small version of the holo into the air in the front of the car. "Some volcano must have gone live there!—Snowden??"

Fireballs soared across the London skies. As if the Londoners were right here, traveling this rutted road up to Hengill with them, their miniature images screamed and fled into the Undergound in terror.

They heard a voiceover say, "But they reconsidered and resurfaced with a proud slogan."

"Can you see it, Freya? A big painted sign... it says: 'We will never give up!' Like in that war." The vid showed Londoners sitting and standing on their front steps, singing, with fires for warmth burning in cauldron-like bowls.

In his mind he saw the gypsies at the Las Cruces Family Campground singing around Malo's fire as loudly as they could, joyful at times and mournful at times. They would be singing now, too, for their own dreams of the future, in the way people sing who know that there are few happy endings, but the song is everything.

**

The shaking ground made their cargo rattle in the rear of the vehicle as they let Norm and Alexis out at the cardboard Festival Personnel sign. The two of them tugged their packs out of the back and disappeared, running up the dirt road into the dark.

Diana got the vehicle parked in the dirt lot with the hand-lettered Sciences and Media sign. As they opened the doors, Ozzie heard Norm's voice coming through the Festival sound system already, calling out over the rumblings to the valley full of people:

"Good morning, if that's what this is, and welcome to Radio Reykjavik, presenting Reykjavik Rocks! Where you are the stars!"

The valley shook nonstop right now, about two inches in each direction. It was an icy cold pre-dawn, and it would be pre-dawn for hours; the dark bowl of the valley was lit with thousands of flashlights. Most of the volunteer singers had slept right there, and some had left and tried to return, only to find their spaces filled. Some had never slept at all.

"Welcome to all of you. Thank you for being here. You've heard the news about Representative B. Arnold Raker's confession, I know. But for those of you who were in the portable toilets or on the phone, I'll tell you again!" And Norm gleefully repeated the story. Ozzie could hear the grin in his voice.

"One good thing about our being here round the clock," Ozzie said.

"Yeah?" Freya hugged his arm, sharing warmth.

"It helps Norm to do more hours on the air. Hope it doesn't go to his head."

"Too late to hope. Listen to him." She chuckled.

"Folks, here's the next exciting episode: Raker has now been served a warrant issued personally by the President of the Oklahoma Tribal Council, for Tampering with the Earth! What do you say, people? Let's hear it from you tree-huggers out there!"

They roared.

"OK, thank you, thank you, don't wear out your voices! let's warm this place up some more! Let's sing the Martian Growing Song again."

CHAPTER FORTY-THREE

FOR MILES THE PLACE WAS SO JAMMED with volunteers and vehicles that no one could go anywhere for food. New since last night, a group of people in matching yellow jackets had formed a mile-long bucket brigade from outside Hengill, passing ten-pound food packets down into the valley full of people and laterally to all the singing audiences on the rim. The bucket brigade sang heartily too, aided by monitors that showed the words and the notes.

He and Freya stepped up to the line and took a packet for them to share with Diana. A man managing the line said, "Hey, you two are Festival personnel, right? Work with that lady scientist? Take these, one for each."

The three extra packets turned out to be heat-protective suits. Freya smiled at how much they looked like her firefighter gear, only these were the latest: as light as nothing at all. Ozzie shrugged and put his on, experimentally. Then, without asking, he dressed her in hers. He filled the water bottles and hung them from her belt and his.

They went to give Diana the hot coffee and food and a suit. After she had collected measurements from her instruments, they helped her stow them all safely in the vehicle and left her there to sleep. The three of them would continue to take shifts sleeping in the oversized cargo compartment.

For now it was his job and Freya's to stay awake and sing, so they began to walk the rim of the valley, soaking up the music and getting warm. Norm called out, "Rest time for all you singers out there! While the next band sets up, I want you to hear the winning Growing Song Contest entry for the nation of Malta; this is 100,000 people singing—believe it or not, a fifth of the population—who got together on a convenient hillside somewhere to belt out the song. Hope they're still out there singing, don't you?"

Freya began to sing softly with the recording, as if to herself at first, then louder. She had a pretty good voice. He listened a little, then sang too. When the band started up and the singing in the valley around them rose again, they gave the song lots of volume, making up in energy for any faults in their music.

**

For hours now the valley had been singing almost nonstop, with all the bands doing nothing but the

Growing Song one way or another. Diana was awake again. Time for Ozzie to go sleep. He didn't want to leave Freya, but she promised to stay with Diana. He looked at the two of them, shining silver in their heat-suits, with the cowls thrown back and knit hats pulled down over their hair. *Both safe.* It was more of a hope that he was stating to himself than anything else.

Together the three of them paused to look together out over the valley. Phone-lights and candles made it look like a wheelbarrel full of piled-up stars. He thought of sailing through space someday, with Freya, through stars almost this thick...

"Ozzie, look!" she clutched his arm.

Diana pulled out enoculars and held them to her eyes. A thin, steaming crack had appeared in the valley floor. "Right along the fault line!" Diana said.

As if it were one organism, the audience down there scrambled to its feet and plunged screaming to either side, away from the smoking line that was growing longer, like a gray serpent, in the lights from the stage. The band sounds died out.

"Norm," Ozzie whispered.

Norm's voice broke the silence, calling out clearly over the panic of the crowd: "Folks, WALK as fast as you can, uphill. WALK, NOW. And don't forget to keep singing!" he laughed. "You guys on the hillside, still sitting there holding down your turf—that spot is no good to you if this place blows, so MOVE NOW, move, and make room for your neighbors to move uphill behind you. Keep on singing!"

Their song roared again along with his, filling the valley. *Just voices, but so much voice power.* The thousands of lights swarmed uphill, slowing on the steep parts. The singing grew even louder.

Norm's radio voice rose above that, calling, "Hit it now, bands! All of you, help those singers! And everyone, migrate uphill NOW!"

The music started up again, raggedly. Ozzie tore his eyes away, to find Freya and Diana still beside him, still mesmerized. He looked again at the steaming crack below. "Let's get up to the rim," Diana said.

The valleyful of singers roared. Norm repeated his message every twenty seconds or so, like a chant he had made up: "Keep on singing, move uphill, help your neighbors, all of you, get on up there, don't stop singing, move uphill…"

They swarmed, stumbled, and defied gravity, flowing uphill.

"And now, Ladees and Gentlemen…"

Ozzie turned. He was frightened, but he couldn't say he was surprised: Ilse and his father ran onto the stage, waving and smiling, singing as they came.

Ilse and Doug had probably been wakened from their only sleep in the last 24 hours. And where was their band? Among the ones that flew out yesterday morning?

The crack hadn't reached as high as the stage yet. But

it moved upward slowly, closer every minute.

**

"Ozzie."

Daylight had arrived. They had never slept. He and Freya stood at the lip of the valley, reaching a hand to each new person who arrived at the edge so the crowd could move uphill fast. Diana took new measurements, on the wooden platform about ten yards away.

Freya squeezed his arm again and pointed: Ahanith had arrived on the stage and joined Doug and Ilse in her song.

He gazed. Her beauty was astounding. He could feel it like a wave. Doug and Ilse harmonized while she took the notes of the song all over the place, Memphis blues style—or something like it. It gave him chills.

There was a swelling gasp and cry that spread around the valley as Ahanith's form gradually became visible to the Festival guests. Then the yells grew into a roar when some of them saw the narrowing crack and the steam that was subsiding.

"That's right, friends, you have help here! But if you stop, she can't do it alone. Keep on singing, move on up."

Some people turned and plodded uphill dutifully but many froze and stared. "OK, you may think the fireworks are over but don't leave it to this lady down here to keep your butts out of trouble. Keep on SINGING! Get on uphill NOW!"

The people walked, ran, panted, ran, sang their way uphill. Someone stopped to try to take holos of the singer.

"MOVE!" Norm yelled.

Despite everything, the crack opened again, and more steam rose. Ahanith sang louder.

"Keep going!" Norm yelled. "AND SING!"

**

Now volunteers rode all-terrain vehicles toward the midpoint on the hillside where the stage rested. They went to work to loosen it from the pilings and steel that tethered it there, but when the crack neared the foundation of the stage, they gave up and hastily dragged some of the speakers and stage lights uphill on something that looked like tekryl sleds, to throw together a makeshift slab of boards and boulders at the rim. Ilse and Doug and Ahanith continued to sing as they hiked up the hill to the new stage.

At this side of the valley, Ozzie saw the media trucks moving uphill behind him.

The crowd, too, had mostly moved to places above Ozzie, Freya and Diana: they were hundreds of yards deep all around the rim of the valley, singing, and up on higher ground above the rim, on all sides. Tens of thousands of them formed a dark collar like vegetation, still singing.

**

Night fell again quickly but not one of the three of them dared to sleep now, except briefly, sitting on the hillside with their arms and heads resting on their upraised knees and one of them at watch. Diana dozed, woke, went and returned, taking her seismic measurements again and again.

In the darkness the crack was a lighted wound like a raw red scratch across the depths of the valley bowl. The singing continued. Ilse and Doug and Ahanith left the stage for breaks, but they returned too soon to get any sleep, Ozzie thought.

Sometime in the middle of the long darkness there was a gigantic groan and Freya shook him awake. The slash that split the valley began to widen. Screams of horror fell in waves around them from the crowds above, as the slash grew wide enough to see red lava move inside the wound far below. Ozzie stood up numbly. Freya too, and she was crushing his hand.

Then the wound ripped wider and longer, toward both ends of the valley. The valley was splitting in two way down there, and the split became a gap that you could look through for a glimpse into hell.

"Keep on singing! Move up higher!" From the stage over there at the rim of the valley, Norm chanted again to the stragglers in the crowd. The other bands seemed to have gone uphill with the media; Ilse and Doug and Ahanith sang on, though. The end of the crack had crept to just the length of a trading ship away from the three singers and the stage.—Just as far away from them as the length of the *Liberty*, which drifted now in the quiet of space towards Mars.

They were here, though, stuck in this nightmare. The crowd roared out the song again. The crack was too close to the stage. *Norm, say goodnight and pack up. Get them out of there.*

Diana had to grab them both by the arms and pull them uphill with her—away from the end of the crack that inched upward on their side, too, steadily toward them. The crowd at this part of the rim had already cleared out. He saw the gleam of Diana's equipment as it bobbed uphill way ahead of them, moving fast on the shoulders of some people she had commandeered in the cause of science.

**

Ozzie turned again toward the valley. "Dad, Ilse," he yelled. "Get out of there!"

Of course his voice wasn't loud enough. "NORM! SINGERS MOVE!" he bellowed. It still wasn't enough. Freya recovered her wits and texted Norm. Diana phoned. Someone with a holophone public address feature took up the call, and others joined right after: "SINGERS! MOVE OUT!"

Ozzie pulled up his enoculars, panting with panic. He saw Norm walk to center stage, grab the mike to interrupt. "Singers MOVE UPHILL!" Norm yelled. Ilse and

Doug and Ahanith kept on singing. "MOVE UPHILL SINGERS!!" Norm stood on the makeshift stage, staring at them. Alexis was beside him, off-mike, her mouth moving too.

Ozzie held his breath. *Go, go, Norm,* he begged silently. *Dad, Ilse. GO.*

Norm abruptly grabbed Alexis' hand and ran out of the floodlights, into the darkness. Ozzie hit the Night button on the enoculars, exhaling with relief. *Look at them go!* She was right with him, scrambling uphill and away from the crack, not missing a step.

But Dad and Ilse! They kept on singing, with Ahanith between them, their faces lit with happiness. He saw Norm turn from the grassy hillside to call down to them, then run upward again, Alexis pacing him all the way.

Ozzie looked down at the crack on this side. It was already past the rim here, and nearing them again; a strong fella had grabbed his mother's arm and was half-dragging, half-carrying her steeply uphill, at a right angle to the split. "That way!" Diana pointed. He and Freya followed her, running upward.

But the two paused to pant, and he couldn't help it: his eyes were dragged down into the valley again. The red crack boiled over now. There was no stopping it. The sight hypnotized them both: a bubbling flood of liquid fire quickly filled the bottom of the valley, then the violent red lake began to rise slowly upward on the lower valley walls, turning slabs of snow to steam, licking the drying grasses with flame—

There, at eye level across the valley: Ozzie's stomach went heavy with horror. Ilse and Doug had not moved, even now! Only the tekryl sleds, now their platform, protected them. Dad and Ilse and Ahanith sang as the crack widened beneath them.

Lost in their song.

Ozzie looked behind him at the pre-dawn eastern side of the rise, where the crowd struggled further uphill, singing. All around the valley, they sang at a nonstop roar.

He and Freya turned toward the crack they had been fleeing. It had extended uphill to split the earth. Still thin as a red thread there, it separated them from the turf beyond. In the dim light he could glimpse the red of lava through the sneering crack, a window to violence so huge it was impossible.

Beyond it and below, three voices still led the crowd: Doug and Ilse and Ahanith.

Freya's hand clenched his. Now or never. He felt her answer. Without a word, they pelted forward, leaped the crack and ran hard, heading around the rim of the amphitheater toward Dad and Ilse.

"Ozzie!" he heard his mother shout twice in dismay behind them. And she was right. Every space-disaster lesson he had ever drilled at the Academy was ringing in his ears right now: Save the many, not the few. Leave behind the ones who won't move. Survive to create safety for others. *Norm was right to take Alexis and run away from the danger. But we are running toward it. We are both insane.*

None of that mattered. He knew Diana was going to be OK. They had to get the others.

**

Lava boiled up in the slash under the stage, across the valley from them. Ozzie could see it as he ran. Now it was a river of lava, widening beneath the three singers. He moaned. They ran on.

The crowd on their side of the crack was running again, screaming and not singing, in every direction that was uphill and away from the lava and the little temporary stage.

The stage blazed directly ahead of them now, soccer fields away. Their running feet pounded across the hillside toward it. The brightly-lit platform twisted and lost its balance, rocking on the rim of the valley. How could they keep playing? But they did. And they sang.

The crowd roared the song, and it drummed in Ozzie's ears.

BLAM! Once. Then twice. Three huge explosions blew—

As if it were a dream, in slow motion, Ozzie saw the stage swell upward with Ilse and Dad—and Ahanith—riding a tide of lava on a tekryl sled like people surfing the enormous wave of a tsunami. There the valley rim rippled and split, spewing lava and fire straight up. Lava threw the raft skyward with the three singers still on it, still singing, among comets of flying molten stuff.

And then Norm and Alexis appeared beyond them!—flying upward on a giant slab of turf, with gigantic clods of earth and grass airborne all around it in the dawn light.

"NOOOOOO!" He and Freya both yelled in horror and despair. They turned a hard left and struggled up the hillside nearest them again, away from the destination they had been crazy enough to try to reach. Their destination disappeared in fire and melting rock.

They were far behind the rest of the running crowd, far from Diana—who was much safer, thank heaven, and still on the ground somewhere, not airborne.

They were too horrified to reason or grieve. They ran.

BLAM! Another explosion. Now they flew through the air too, high above everything—fire and lava and the screaming crowd—on a huge hunk of earth, big as a soccer field, like a spaceship hurtling skyward. Lava spewed skyward all around them, too.

The hills roared. They heard Ahanith's voice. Somewhere in the distance the crowds sang. But everything near them seemed to stop and float.

They had ceased rising. They hung in the air for what seemed like forever. Then with a sickening lurch in his stomach, Ozzie knew they were dropping.

They fell a long crazy way downward. As they fell Ozzie ran toward her, and she began to run too.

She could feel the impact coming close beneath them. She stopped, holding her breath, waiting to be hit. Instead, they stopped plunging earthward with a bounce that knocked them off their feet, but pretty softly. She sat up and looked around with relief.

But now she saw what had arrested their fall: lava rising around them. Heat poured toward them as fire dried and lit the grass at the edges of their earth boat. It had caught up with their huge hunk of turf now and the tide lifted them back upward again slowly, like the swell of a wave in the North Sea. A wave of liquid fire and rock, she thought numbly, terrified.

Why wasn't the heat she felt worse than this? *The heat suit,* she thought.

Both of them scrambled to their feet, starting toward each other again. Ozzie had fallen wrong or something. He was only twenty meters away, but he was limping so badly it slowed him.

The heat suits: she recalled emergency training and tore the water bottle from her waist. "Ozzie!" She yelled to him, doing hers with large motions for him to see: poured the water over her hair, soaking her head and drinking the last mouthfuls before she tossed the bottle and cowled herself. She pulled on the heat mitts. He nodded and did the same, then limped forward again.

Her heart stopped. Without warning, the ground under Ozzie split away, becoming a littler island of earth about ten meters across. It began to heave up and down terrifyingly on the roiling tide of lava. Ozzie lost his balance again and went down on his knees.

On the larger island, which was at least rising steadily now, Freya ran toward the gap to help him leap across the meter of lava flowing between.

He screamed at her to go back and leave him, find safety! *Safety? Where?* She reached into her heat suit at the neck and tore out her Egyptian scarf. She ordered him to come grab it, praying for luck. The gap narrowed a little. But the lava at her feet made her sick with fear, pouring heat up at her. She swung the brilliant scarf for him to catch the end. The gap was widening, closing, widening...

In her mind she saw the heroic cats in the Mars food capsule who had died to protect a litter of future cats. *Would I be brave enough to do that?* she had wondered.

Of course, she answered furiously now, taking her own dare. She ran and leaped across the lava stream, landing pretty neatly for a leap to such a tipsy raft of dirt, and tugged Ozzie back toward the stream to help him jump across. But her leap had widened the gap again. She begged for it to close.

Then she knew that only one of them would make it; one jump would probably send the smaller island skating away on flaming melted rock. She knew, stubbornly, that she had to get him across, for all the future that he had dreamed. When it narrowed, so the two islands were nearly touching, she was ready. She shoved him ahead of her.

He pushed back, away from the edge, and grabbed at her so she wouldn't fall. They struggled together beside the gap, by flaming grass, boiling rock—and lost their balance. "Ozzie!" she screamed in despair. They fell to the ground on the smaller island, and it slid away from the big one.

Horror paralyzed her. She stared. The capricious lava upended and swallowed the entire larger island, the size of a soccer field, in one long nightmare gulp.

She had almost doomed him.

They struggled to stand again together, arms around each other, on their little hopeless bit of land. There was lava-fire all around them, and where it was close it was as blindingly bright as looking into a candle flame. Impossible.

Then, even more impossible, she heard her mother, Ahanith and Doug singing, a towering crescendo that rose higher and higher above the roar of the flames and boiling rock.

An explosion engulfed everything.

Pain stabbed fire through Freya's ears. Her scarf rippled away like a flag. They fell again to the ground and their island shot upward on a crest of lava, flying as if gravity had disappeared. Molten spray flew too, all around them. Ozzie rolled on top of her to protect her. *Don't get burned, Ozzie,* she thought, hugging him to her with all her might.

This is it. We tried to save each other, and we did. But there's no saving anyone here... she clutched him tighter, as if that would protect him too.

They began to fall fast again, at a sickening pace, faster as they went, and it lasted forever...

SLAM, they slapped down, like being body-slammed by a giant. They rolled together to a stop. Then she couldn't stay awake any longer.

There were choruses of screams. Across the green hillsides people ran, chased by rivers of hot rock and geysers of steam. Sun shone on them, rising sun. They ran and sang, ran and sang, and above their singing she could hear the Singer's voice...

She saw that people were standing in the streets of Paris, singing and walking, like a funeral march, on the Champs Elysee. In gray light and swirling snow, the Eiffel Tower was washed by thousands of waving flashlights.

Through swirling snow she saw a horseman deliver a folded sheet of red stuff to a girl at the entrance of a large, snow-drifted yurt. When the longhaired girl opened it and it sang to her, she carried it inside and stood beside the pulsing fire and steaming pot, to have the others sing with it too.

People crowded the boats hung with lights off the

coast of Sicily, holding more lights, singing as their long procession wound around a harbor.

In northern Australia, people held hands in a chain that extended along the coastline, like a protective cord. There were lights there, too. They sang.

People were singing in Botswana, loudly and fervently, swaying together.

She could hear them all somehow, as if it was a dream.

It seemed that she and Ozzie were flying, seeing it all, like that vision in the house with the ibis door in Egypt. She didn't know where she was...

She didn't know where she was but she remembered the rules: if she got lost, she must find him.

He appeared. She took his hand.

Didn't they need to be gone from here? If they needed to, their Egyptian guide had said, they should just decide to be somewhere else...

But she was just deciding to go, when something else caught her attention.

In Pasadena, rioting had turned to singing. And wild dancing in the streets. At Jet Propulsion Labs, a crowd watched instruments and sang.

At a funny long restaurant in Arkansas, a bunch of old men were mumble-singing the Growing Song together, huddled over their coffee and frightened each time the building shook, worried that the world would end. A plump, kind-faced woman stood behind the counter—someone who seemed familiar—wiping her eyes and singing, making them sing with her.

As dawn neared in London—while their elders were wearily sitting, standing, supporting each other, still singing faintly—young men and women had begun to run and leap over the street-fires, daring the fire to lick at them, daring fire to vanquish them. They were wild-eyed and red-cheeked, breathing hard in the soot and smoke, singing and looking at each other with the sudden hunger of those whose lives may be short.

Ozzie still had her hand. She heard him talking to her. Everything seemed to be shaking around them, the ground under her rumpling and straightening like a shook blanket.

It was time to leave for someplace safe, so she looked at him to change the vision. But he didn't seem to see her. He didn't look.

**

Had a long time passed? Now she heard wild howling. A wolf crying out at the sky, but there was no moon... The ground rumpled and shook, but less, it seemed.

Freya woke.

Ozzie was kneeling beside her—crying? Why?

Because our parents are dead. She remembered. *Could that really be true? Was it a dream?* She scrambled awkwardly to her feet to look around.

Ozzie stared up at her and stumbled to his feet, too. She had never seen him look so surprised.

**

She's alive.

He disbelieved it. He took her hands: still cold. The earth shuddered again and again around them. Blood was still oozing out of her ears, staining the wisps of light hair.

The blood was from her eardrums?—broken by the explosion?

He stared at her face, oiled with sweat and soot. They were standing on ashes in the heavy, sulfury-smelling air, below low, dark-gray clouds that seemed to be swallowing up the rising clouds of steam. But what was amazing, he realized, was just that: they were standing.

Surrounding them it was as desolate as if a bomb had been dropped, desolate as Mars—but not quite: warm rain had begun to pour down on them and all around them.

"We made it!" he yelled right in her face. She smiled a little, and nodded numbly, clumsily. Seeing her blood made his heart ache. Fury blinded him.

Ozzie never could remember, later, what it was exactly that came over him, but one thing led to another.

First, when he folded her in his arms, wanting to care for her so much that she would never be hurt again, he started to bawl. With tears running down both their faces, they hugged each other and sobbed like two forlorn children.

Then some wildly raging joy and triumph took over him; it made him kiss her hard, then harder, then again. We have won this battle, and we will never give up! She kissed him back, just as crazy hard.

It was only when they paused, stunned, to take a breath, and they looked in each other's eyes again, that they began to laugh. They laughed with relief and delight and joy. They laughed even more than they had cried, till new tears ran fresh tracks down the rainy soot on their faces and their legs went so weak they sank down onto the wet ash, pounding the ground with their hands.

They were still laughing when Norm and Alexis caught up to them there.

CHAPTER FORTY-FOUR

OZZIE AND FREYA DIDN'T EVEN GREET the other two; they just laughed harder.

"Glad to see you too, Ozzie," Norm said. Freya staggered to her feet again and hugged them both fiercely. Ozzie rose just in time to lose his balance in an aftershock. He fell laughing into Norm's arms.

Then he noticed the blood on Norm's ears too. "Hey! Norm!—"

"That's what I call a ferocious hug, Freya," Norm said very loudly.

"Adrenaline," she laughed, facing up into the rain and swiping the soot away from her face with her hands. Then she stopped and stared. "Last time we saw you two you were—"

"Yes!" Alexis said. "Freya, you won't believe what happened!" The fishtail braid was gone: her hair was burned off at her shoulders, ragged and fuzzy.

"Try us." Ozzie looked at Freya and grinned, then

eyed Norm, then became suddenly, overwhelmingly sleepy. He had been here a couple of lifetimes—a few hours, at least—watching over Freya, most of one short winter day. It was dusk again. The rain still fell.

He was sopping wet. He ran a hand through his hair, as short as Norm's now. The arms and legs of Freya's heatproof suit had been evaporated to a few black shreds hanging from her bare skin; and when he looked his was the same.

He sagged to sit on a rock, and Freya took the other end. An aftershock almost shook them off it. Freya chuckled.

"First, look at this, Ozzie." Norm held out his phone.

Ozzie held out a hand with his eyes closed. "Resting, Norm. Justaminute."

Norm couldn't wait. "Raker got busted again! He had an escape ship set up for him and friends to leave from New Mexico, in case Earth ever looked risky, for his 'summer place' on the Moon. And we know what he had in mind once he got there.

"You will love this. While he was out on bail, he headed for New Mexico to escape at the Spaceport. Somehow word got out and his escape ship was grounded by an Oklahoma gypsy posse that showed up. He has been taken to Oklahoma for trial."

"No bail this time, so sorry," Alexis scowled. The ground shook again and she fell over, giggling.

Ozzie and Freya both lost their seats on the rock and

laughed all over again. So funny. There they were, lying in the mud and ashes, laughing.

"Dark coming," Norm said. He ran his eyes over their rags. "Since you two are dressed for the tropics, it could get a little chilly whenever this rain turns cold. Let's move. We can tell our tales as we go, buddies." He called someone using his phone. "No phones, guys?" he asked Ozzie and Freya.

"Gone," Ozzie said.

"Mine's drowned," Alexis commiserated. "Gone too."

"Mine is ultra-waterproofed," Norm said smugly.

"Your dad and Ilse..." Alexis began.

Freya searched her face and Norm's, then just nodded.

Ozzie put his arm around her shoulders. "If we've survived, who knows who else did, right?" *I sound like Dad.* "We'd better go find out. I don't even know where we've ended up. Do you?"

Alexis pointed. "Let's go," she said, and led the way. Ozzie knew that it would take concentration just to follow her, right now.

Ozzie and Freya limped uphill and across the turf, with help here and there from Alexis and Norm, till a hover-vehicle met them on a small road somewhere in the darkness. By then the aftershocks and tremors had dwindled.

Everything smelled burnt. Warm rain still fell. Through a haze of weariness, Ozzie saw that Diana was there too. Her eyes were like an embrace. Ozzie asked: No, no sign of his father or Ilse.

Hands put them into the back of the ambulance, dried them off, belted them into shelves. Gave them first aid, urged them to sleep.

**

But Norm wasn't going to be shushed. He told their wild tale: thrown skyward, their raft of turf had landed high on a steep gravelly hillside and slid downhill till it disintegrated, leaving them rolling toward a thermal spring with Alexis' braid on fire... "I had to dunk her to put it out," Norm chuckled.

"Pffff!" she said. "The spring was about ten feet deep, friends! Any less and we would've been as beaten-up as you are! And HOT in that water! Stunk like eggs! But we were so cold from being out in the wind all day that it felt good, for about five seconds. We were lucky it wasn't one of the *boiling* ones! But scary as hell, literally, so we got out of there as fast as we could and dove into some snow to cool off, and ran—"

"—But we were far enough from the lava flow by then—"

"—So we found people, but not you, and left word—"

"—We asked, and Diana was fine they said—"

"—And came to find you. We were scared. We were scared about you two."

Now their vehicle was gliding above the countryside

toward Rekjavik. Ozzie let Freya tell their part of the tale. It was nightmarish in his own memory, so he liked hearing it told in her voice with his eyes closed. Maybe telling it made it less nightmarish to her, he thought. By the time she finished, her voice was smiling and the lights and noises outside told him they had entered the city.

"Hey, why didn't ya check her heart, Ozzie? Then you would have known…"

"Sure, Norm. The ground was pounding and shaking at the time. Lub-dub, who could hear?"

Norm changed the subject hastily. "Hey, where's Ahanith? Do you know?"

They all were silent.

Ozzie feared the worst.

**

It was after midnight. The search for their parents, by some of the hardier and less frightened Festival guests, had continued for hours, and would resume by plane and copter as soon as the sun rose; but by now Freya knew.

They had all sent messages, to everyone who cared, to say that they were OK. They sat in the kitchen of Freya's house wearing anything they could find that was dry, nursing cups of tea or soup in weird celebration of being alive at the end of this nightmare day.

Diana sat silently, nodding or smiling a little at each

of them now and then. They all chatted, told details, asked more questions of her and each other. She seemed to know that they needed this.

Diana said quietly, "PII's seismic timeline shows that the final blow at Hengill 'let off steam' for the whole planet by allowing the worst area to settle, and after that there was settling everywhere. It will take a week or more, they say, but right now the whole planet is moving closer to normal vibration and volcanic activity as it was last recorded three or four years ago. It's remarkable."

They all would have shown more enthusiasm, normally.

Freya looked at her mother's herbs, vibrating only slightly on the windowsill. She didn't want to think about it. And they still had no word from Ahanith, or about her.

Ozzie looked like he had gone into shock. Freya knew the signs from firefighter training, and also knew that her own shock would come on a delay. It always seemed to. As Diana talked, Ozzie looked at his mother now and then with relief. And she wore the same look when her eyes rested on him sometimes. Or so it seemed to Freya. If it was relief, she understood.

**

Freya woke again hours after they had all gone to bed. Terror churned in her stomach, and as she realized who and where she was, her terror was replaced by the numbness of a shock so deep it held her awake and desolate while the others slept. Her head pounded.

Then she thought she heard Ahanith's voice. From under the covers, Freya could hear her singing softly somewhere in the house. The music crept into her ears and landed on her hair, her eyelids, her skin, like sunlight or warm wind...

When she woke again much later, the voice was silent.

On the newsfeeds some of those who had witnessed Hengill's "big blow" told reporters that they had seen a filmy figure with a crown-like hat on the stage singing. But no one could produce a photo as proof.

Long after noon, when they all were finally awake, Norm received a holotext from Malo. "'Ahanith is with me, all safe!'" he read to the other three.

"Whaaat? How did she get there?" Alexis said.

Freya and Ozzie looked at each other wonderingly. "No jar!" Ozzie shrugged.

**

They all seemed to just want to be together. No one could stand to be off somewhere else; even Diana, who had to pack, kept hovering in the kitchen.

Someone had noticed that—Alexis?—and threw an old 1000-piece jigsaw puzzle down onto the kitchen table to give them something to do together. Alexis and Ozzie sat sorting the pieces. Ozzie looked a lot better than last night.

"I've filed that suit in Icelandic court. By phone," Norm said.

No one was listening, so he said, "The suit against Raker for hiring boats to attack us in the North Sea."

Freya stopped drying the dish he had washed and gave him a thumbs-up. "Good, Norm. He deserves it."

He washed another plate and picked up his phone again. Dishwashing was slow this way, but at least he was doing it.

"Can you believe the nerve of that guy?" Norm said to his phone. No one paid any attention, so he added, "Seth..." That got their ears. "Seth says, 'You wouldn't dare press charges against my father.'"

Norm punched a couple of buttons and used a souvenir Hengill photo from his phone as background. That way he could stand here in the kitchen and to Seth he would be recording a live holo reply in front of some Hengill volcanic steam. "Quiet, guys," he hissed.

To Seth he said: "You *betcha* we would." He clicked off and sent. Ozzie grinned at him.

"You betcha?" Freya turned to Alexis.

"It means 'you bet ya,'" Alexis said. "Like, 'you can bet your life on that.' True, Norm?"

"Yeah," he glowered.

"Okay: you betcha," Freya chuckled. Then, softly, "You betcha, Seth. You betcha, Raker."

**

There was a long crack, from floor to ceiling, in Freya's bedroom wall near her bed, so thin no one else would ever notice. Only she, who had memorized every bit of that wall, would know. In her own thought, she was sure: the crack must have happened when Hengill blew.

**

Freya received a medal for heroism from the Icelandic government, and a festival in the harbor, a week later. Ozzie instigated it, but it wasn't hard to arrange: the whole thing exactly suited the mood of the city at the time. Reykjavik was crowded with international visitors who were reluctant to leave the scene of their triumph just yet. And some emigrants were already returning home again from Sweden and Norway to help salvage their homes and clean up the place—including someone, Freya was sure, who could replace Ilse as an art teacher. She said it looked to her like a transfusion of people and ideas for Iceland. And a celebration was just the right thing, to all of them.

Colored lights were strung across the harbor, and the 14th Reykjavik Fire Company contributed arcing water-gun salutes. Norm and Alexis stayed for the ceremony; they wouldn't miss it, they said; but it was a good thing because they also received medals, along with Ozzie's. Ozzie was the only one who was surprised.

Diana and Dr. Nimr returned especially for the event;

Dr. Nimr shook all their hands warmly, and embraced Freya and Ozzie in a fatherly sort of way. After the ceremony, the six of them walked around Old Harbor watching the bonfires, the fireworks and revelers.

Norm couldn't resist saying it, Ozzie guessed: "Hey Freya, your crazy idea about the music of Osiris? It worked!"

"Ja," she said, still in Icelandic mode from her acceptance speech. "With your help. And Ahanith's. With huge help from you all." She looked around at them, and her eyes showered blue sparks.

Ozzie said it: "Those holofeeds we're all getting from the world scientific community: The scientists all *know* that it was the big power sources and the correct wavelength that did it—"

"—But the Singer, and Malo and his friends, and everyone who sang here in Reykjavik—all of us know what really happened, what we all did, and what it took." Diana's eyes were steady and blue-gray.

"Fortunately, PII has the seismic documentation to prove it, in case anyone ever needs to," Dr. Nimr added. "Here's something you might like to see." He pulled out his phone and hit a few keys, then waved up a holo that floated at the center of the circle they had formed. It was a sphere overlaid with a net of intersecting lines. Ozzie recognized the landforms and blue oceans of Earth inside the net. The lines began to glow, and to leak light from each knot of the net.

"I've seen that!" Alexis gasped.

"I dreamed it," Norm said.

Ozzie and Freya looked at each other. "We did too," Ozzie said. "What is it?" he asked, and then he thought he knew.

"It's a sort of data-generated schematic map showing the electronics of Earth, 72 hours after the vibration sites were shut down—24 hours after Hengill blew. It shows the harmonious flow and linking of energy that resulted, like the energy patterns on a healthy human body.

"Three days before, at the time when the twelve vibration sites were shut down, this is how it looked"—in the holo he now showed them, the lines were a torn and tangled net, with its links scrambled and clotted—"and during the final three days of the Hengill Festival, when the whole world was singing"—Dr. Nimr showed a time-lapse view: the lines shifted gradually, then very quickly at the end, to became an orderly, pulsing pattern.

"Does the energy net have a sound?" Ozzie asked. But he thought he knew the answer to that, too.

Dr. Nimr brushed a volume icon with a finger, and now the pulsing net sang.

"It sounds that way now?" Freya sounded like she could hardly make her voice work.

He nodded.

"It's Ahanith!" she said. "The Growing Song, the music of Osiris."

**

The Rekjavik kitchen was sunlit and warm, although snow frosted the sills outside.

Ozzie looked around. It looked the same here as the night he had first seen it, except nothing was shaking. Those rattly little pots on the sill sat fat and green and quiet, now. The house was mostly emptied of musical equipment, donated to the school and the museum. And the place was cleaned so the relatives could come from Sweden and Denmark to stay here while they repaired their homes in the city.

As it turned out, Alexis and Norm had remained a couple more weeks to help out. To celebrate, too: Freya insisted that they all cash their little PII paychecks and spend them dancing at some of the cafes and climbing into some glaciers and thermal springs.

All their backpacks were at the door. They held their cups, looking at each other one last time before they separated—but not forever, Ozzie was sure.

Berkford had just offered Norm his scholarship back because of his "outstanding research achievements," and other general statements.

Ozzie wasn't surprised at the offer.

"They said I have at least half the credits I need for my degree already," Norm said. "But let's face it: I'm famous. They have no choice."

"I'll bet somehow you end up owning a radio station,

Norm," Ozzie grinned.

Alexis' sideways smile went up her face. "And…" she prompted Norm.

"And I told them 'no thanks, nice of you all—but nah, first my girlfriend and I will be starting our own company.'"

"What??"

"No official announcements to make yet, folks.—We need a little time!—But we *are* starting our new company right away," he said. "I have a holo-drive with all the Mars greenhouse specs and info, and we will build self-regulating greenhouses like it and become rich as King Tut! or was it Akhenaton. And his wife. What was her name?"

"Nefertiti. With Mars Colony permission—and we think we have a way to get it—we can do greenhouses in Egypt and other desert countries, to get more things growing again!" Alexis was so happy she could hardly contain it, Ozzie could tell. But she wasn't crying, he noticed.

Freya gave them both her congratulations, her eyes waterfalling blue sparks. "How perfect. You will both probably get honorary degrees anyway," she said.

"New houses for our parents!" Alexis giggled.

"Motorboats too," Norm grinned. "Private planes for the siblings…"

"A garden for our kids—" She caught herself, froze, and watched his face.

Long pause, while everyone watched everyone. Ozzie wondered if Norm would drop the ball.

Finally Norm nodded slowly. "Good karma," he said. He put out his arms and hugged her, and broke into a grin over her head at Ozzie and Freya.

Now she *was* crying.

CHAPTER FORTY-FIVE

SPRING HAD ARRIVED EARLY THIS YEAR, in the hills and farmland south of Space City. At the gypsy camp in the Las Cruces Campground, Ozzie and Freya were welcomed with more warmth than ever before. On the day of their return to New Mexico, at a huge breakfast that included all the gypsy families, they feasted at the fire circle outside Malo's tent.

And then they were sung to. Her eardrums had healed pretty well during the last few weeks, Freya thought; she could even hear most of the high notes now.

She was surprised when she heard Ozzie tell Malo it was amazing that people's voices had made such a difference in the earth's crust. Wasn't it kind of a miracle, he asked, that sound, even that much high-quality, live sound, could do such a thing?

Malo said, leaning toward him confidentially: "Singing is powerful! Don't think it isn't. But that's not what's most important."

"What is, then?"

"For them to begin to dream again. Dream of what they want, not what they don't want. Sing about that."

**

The three of them were returning from a walk around the camp when Ozzie said to Malo: "In Giza, when an old merchant talked to us, we all saw visions of the future. So real. But some of them can't happen now." He could recall his own exactly: his father seeing him off for a trading flight.

"You know what I think: you make the future from your dreams. But other people are making it too."

Ozzie thought of what Dad and Ilse had made. Such a music festival. Their idea; he would never have thought of that.

Malo looked at Ozzie and Freya, a long look. "Are you sad that they did what they did?"

"No. Just... sad."

Freya nodded and took Ozzie's arm.

Ozzie heard a mournful gypsy song begin. It rose into the cool morning sunshine from a solo guitar and some voices. They sounded the way he felt.

"They're singing to your father. And your mother, Freya."

When they entered the fire circle again the Singer appeared there too, standing across from Malo. She stood with her head held high: queenly and beautiful enough to make you gasp. Her voice rose to join the others.

**

Malo is an ancient person, Ozzie thought again with amazement. You would never guess. The man stood strong and tall, and announced:

"The Singer has done what she needed to do—for her own salvation and Earth's. She misses her home. It's time for her to go back to Mars and sing there again. She will leave here, with the promise that all our families here will see her again."

He didn't look sad. Ozzie waited.

"PII will deliver her back to Mars. But *I* will take her there," Malo grinned.

Around him Ozzie could see broad smiles. *Probably everyone here knows about this except Freya and me.*

The ancient gypsy explained: "I have a free pass for PII Spaceline Moon and Mars research flights. We will go as tourists now," Malo said. "We can look at some of our favorite old places."

Freya's eyes shot delighted sparks at Ozzie. He knew what she was thinking: how romantic. Strangely, he found that he thought so too.

**

There was something else coming, she knew. Freya turned to watch Malo.

"Ahanith has something to tell you both," he said. He sat down on a stone at the fireside.

The Singer moved to the place before Malo's tent, but she was silent and her image was fading. She seemed reluctant. Most of the families had begun to walk back toward their own tents now, quietly disappearing the way they always did. A dozen men sat and poked at the fire or drank coffee. A hundred cats or so sat in the shadows away from the fire, watching her. *This must be important*, Freya thought, because the attention and the silence from the cats was enormous.

"Better tell them."

More silence. Ozzie looked at Freya. She took his hand, waiting.

The Singer was almost entirely invisible. She began: [I would not hurt you.]

[I know that,] Freya said. [Once you saved my life. More than once.]

Ahanith became more visible. [Yes, I did...] There was a long pause. [They said that they wanted to sing with me always.]

Freya searched in her mind for something Ahanith might mean. Nothing fit.

[I could feel that Earth needed a big explosion to release the disturbance with less damage. Best right there on that old volcano, where the crust was thin. We just had to do it more, not stop, to make it blow up. So we did.]

"Ozzie?" Freya whispered. He nodded, staring at the Singer's eyes.

[I told them to stay with me; just keep singing. It was the only way I saw to do this...] she explained, patiently.

"And they did and the volcano blew and that fixed it?" Ozzie said, like someone reciting a lesson he had just heard but didn't understand yet.

[Yes. They promised that for my help they would come and sing on Mars with me.]

"But now they *can't*." Ozzie seemed to be thinking his hardest.

[I can show you.]

She began to sing her growing song. Her voice was wild and creepy and gorgeous, as always. Freya loved it and it was comforting to her, but—

Faintly, Ilse's voice joined Ahanith's, clear and cool as silver.

And Doug's, rich and warm like gold.

Three-part harmony on wild, creepy and gorgeous— it was their Reykjavik Music Festival version of the Growing Song. Ozzie was trying hard not to cry now. He wasn't the only one. Freya thought she would explode into pieces.

"*Where are they?*" Ozzie demanded of Malo.

She stared at him. His face was heartbreaking.

"Look," Malo said.

And then they saw. Ilse and Doug appeared next to Ahanith wearing cowboy hats and furry vests, waving at them. Ilse wore her delighted smile. Everything is perfect, couldn't be better, the smile said. They leaned toward each other and moved together to the rhythm of the music, singing to each other and Ahanith as if the three were a perfect world together.

The volume rose. Gradually their song included Ozzie, Freya and Malo, then those at the fireside. The lingering gypsies were singing again, too, by then.

When the song finished the three semi-transparent singers bowed.

[Now you see,] Ahanith said. Her eyes were bright again. [They will come to Mars with us and sing it back to life.]

**

When they arrived at Ozzie's house at last and dropped their packs on the kitchen floor, he felt a twist of loss in his stomach at the silence there. But it left when he saw his father in his mind again, singing at the fire circle in the gypsy camp. He shook his head in amazement.

Out of habit he opened the refrigerator. It was full of carrots, beets and onions. He'd never seen their garden produce so much—and this was what remained in spite of all that the gypsy families were eating and sharing.

"Hey, eggs and milk for breakfast," he announced to Freya. Then he looked over at her. She looked back. Her eyes had gone silent and waiting, like the gypsies.

He hugged her, and offered her a bouquet of onions to make her laugh.

**

The mailbox at the front gate was full of offers for Ozzie: to apprentice, to represent products, to make appearances. He'd seen the rest in stacks on the kitchen table.

He looked them over wonderingly. He remembered how scarce these offers were a couple of years ago. Dad would have been amazed. Or maybe not amazed, just happy for him. He knew Freya would be delighted. But it could wait.

There she sat, over on the front step. She was watching two jays squawk at each other on a branch of the cottonwood tree. Her wild hair flew in the warm noontime wind, and she looked like she didn't have one care in the whole world.

**

Freya watched Ozzie shuffle through his mail, loving the way he smiled at some of them. He would tell her about them later. For now it was enough to watch him, loving the bravery of his face. She knew everything about his face now, and loved it all.

Her own mail, in Reykjavik, had been full of interesting surprises:

The 14th Reykjavik Fire company had offered her the

job again if she wanted it.

A European clothing designer asked her to model for a fundraiser to support Volcano Relief. Someone with a line of hair-care products wanted to try them on her and shoot publicity photos. That made Freya laugh.

She had been offered a poetry-reading engagement in Paris next month, which she told them she just might do. Also, a contract to write her memoirs. And a job training cats for a Psychic Circus at Moon Colony Three.

Just now, sitting on this step, she had decided what she wanted to do: all of them. Except for the hair part, of course.

She had already accepted her job with the 14th Reykjavik Fire Company again, a one-year contract to start after the Paris poetry reading. For the next year, her fire company would be cleaning up volcanic damage and restoring the scorched and battered landscape of Iceland, while Ozzie flew around between the planets, apprenticing. It would make the long year away from him seem shorter.

She looked up now to find him standing beside her in his knee-high space boots. Right now his eyes were blue-green-gray, like the sea, or like a star-sprinkled sky at dusk: the most beautiful eyes she had ever seen. Any distance at all from them was suddenly too far away. She patted his nearest booted calf and offered him her hand.

**

His hand went electric, touching hers. He pulled her to her feet and held her wild hair against his face, wanting again to care for her: his loyal friend and fellow traveler, this beguiling person who laughed at him and understood him.

He loved her dreams as much as he loved his own. He loved her radiant witching-wand life. There was a lot that he wanted to say to her about all that.

But he couldn't remember any of it right now. Instead, he rearranged her carefully in his arms so there wasn't an inch of space between them. Then he kissed her.

That ended up taking a long, long time.

**

Diana came by at moonrise. The fat golden-yellow moon haloed her and filled the porch as they opened the door to let her in. The three of them sat together at the table in the kitchen.

"...and you, Ozzie, have you decided what you'll do?" she asked after a while.

"Sure," he said. "I hate to wait, but it sounds like you won't be too long at it. You know what I *want*: I want to captain the first PII trading ship, *Price of Freedom.* I want to take Freya trading with me, and 100 cats to Mars in carriers, and see if Georgie has changed at all.

"So. What I'll do: I'll apprentice with Teffna, GG's closest competitor, for a year, and then fly with them till

PII is ready. Teffna will understand that I'm going to leave to join PII Aerospace after I've given them their money's worth."

She smiled. "Good," she said.

She breathed deeply. "Now that you've decided, would you like to hear what's in the ring?"

"You found out! Why didn't you tell me?"

"I didn't want to influence your decision," she said. "Ready?

"The dark wire inside the ring is what my first tests showed it was: carbon nanotube technology. It has some very interesting properties. But it's taken a while for PII to reverse-engineer what Dr. Tersey had there, and what he didn't, to finally get what he was doing.

"For example, the technology seems to absorb electromagnetic frequencies and can channel them away so they become less destructive. The ring is getting hotter because it's a circle, absorbing more and more over time without the energy dissipating as fast as it's absorbing. The extreme magnetism we saw has something to do with that. The ability of this stuff to channel energy is so strong."

"The carbon nanotube core should not heat up; at least it was not predicted to... We knew about the strong magnetic properties of a nanotube in this sort of ring-shaped configuration, called a nanotorus—and this one is a gigantic one—but heat? That was confusing.

"The heat turns out to be a flaw in the technology

used in this ring. Nanotube technology permits you to contain and direct energy. The ring was collecting heat but due to the flaw it was unable to either contain it completely or discharge it completely. Today that could easily be corrected.

"But trying to find the cause for the heat made us look closer into the reasoning of the person who created the ring, your grandfather. What was he trying to do and how?

"Last week, we finally made a breakthrough and the result, just yesterday, is new discovery! Some remarkable technology is involved here, something that was far ahead of its time twenty years ago. Ozzie, listen to this: Dr. Tersey's ring turns out to be a recipe for a new drive!"

Diana actually laughed to see his shock. "See, you take the terrifying power of the formula engraved on the ring, and add the advanced nanotube technology inside the ring for the strength and control to direct all that force, and you can have a drive at a whole new level of power.—Like a huge slingshot, that makes not only Jupiter and Saturn accessible, but also Vega and Sirius, the far edges of this galaxy, and beyond..."

Her mild eyes were no longer mild. They shone.

Freya stared at her eyes, now just like Ozzie's, which had grown starry too.

Ozzie turned to Freya and took her hand. "Let's go," he whispered, grinning like a gypsy. "To Vega. Will you

come with me?"
 Love talk from a space-trader.
 "You betcha," she whispered back.

***

By J.K. Stephens

The Ibis Door, Dreamers Book 1

The Singer, Dreamers Book 2

In the Ring, Dreamers Book 3

Connect

I enjoy staying in touch with readers, so please join my mailing list for blogs and periodic updates. Just email permission to:

jKStephens@daybreakcreate.com

Your opinion helps other readers

If you enjoyed *In The Ring*, you can leave a short review on the page where you bought this book. Your reviews help readers like you to find books they will enjoy—including this series, I hope.

ABOUT THE AUTHOR

J.K. Stephens lives and writes near Tampa, Florida.

A fan of funny movies, dancing, long walks and bicycling, the author is as crazy about mountains as beaches, and travels at any opportunity—not only to gather ideas for further writing, but also to enjoy the remarkable people, places, foods and dreams that make the world the way it is.

ACKNOWLEDGEMENTS

Grateful thanks to my first readers and beta readers, to fellow writers who have helped, and to friends who have encouraged—particularly Kathy, Janet, Rebecca, Gene, Alan, Marlene, Cheri, Bill, Scott, Valorie, Gayla, Daphna and Tony.

The Dreamers series is made out of dreams, stories within stories, journeys within journeys: it has been rewarding to take all of them, and the overall story, to their destinations. Thanks especially to you, my readers, for listening while I tell this story.

J.K. Stephens

Made in the USA
Columbia, SC
23 March 2025